T0137466

Springer Proceedings in Earth and Environmental Sciences

Series Editor

Natalia S. Bezaeva, The Moscow Area, Russia

The series Springer Proceedings in Earth and Environmental Sciences publishes proceedings from scholarly meetings and workshops on all topics related to Environmental and Earth Sciences and related sciences. This series constitutes a comprehensive up-to-date source of reference on a field or subfield of relevance in Earth and Environmental Sciences. In addition to an overall evaluation of the interest, scientific quality, and timeliness of each proposal at the hands of the publisher, individual contributions are all refereed to the high quality standards of leading journals in the field. Thus, this series provides the research community with well-edited, authoritative reports on developments in the most exciting areas of environmental sciences, earth sciences and related fields.

More information about this series at http://www.springer.com/series/16067

Abdullkhay A. Zhamaletdinov ·
Yury L. Rebetsky

Editors

The Study of Continental Lithosphere Electrical Conductivity, Temperature and Rheology

Springer

Editors
Abdullkhay A. Zhamaletdinov
St. Petersburg Brunch of IZMIRAN
St. Petersburg, Russia

Yury L. Rebetsky
Schmidt Institute of Physics of the Earth
of the Russian Academy of Sciences
Moscow, Russia

ISSN 2524-342X   ISSN 2524-3438   (electronic)
Springer Proceedings in Earth and Environmental Sciences
ISBN 978-3-030-35908-9   ISBN 978-3-030-35906-5   (eBook)
https://doi.org/10.1007/978-3-030-35906-5

This Springer imprint is published by the registered company Springer Nature Switzerland AG
The registered company address is: Gewerbestrasse 11, 6330 Cham, Switzerland

Devoted to memory of Prof. Aida Kovtun

Review

To Proceedings of the 2nd All-Russian (with international participation) scientific-practical seminar "The Study of Continental Lithosphere Electrical Conductivity, Temperature and Rheology". Eds. Dr. of Sc. A. A. Zhamaletdinov and Dr. of Sc. Yu. L. Rebetsky.

The presented collection is divided thematically into two sections devoted to theoretical studies and the results of experimental observations. The total number of articles is 13. Despite the fact that studies of the earth's crust of the Kola region for all its thickness using deep electromagnetic studies were actively conducted in previous years (1974–1990), the issues of studying the structure and nature of geophysical boundaries still remain relevant for the continental crust.

The proceedings also present works relating to the study of the earth's crust in other regions of Russia: the Voronezh Massif, the Yamalo-Nenets Autonomous District. Distant (foreign) regions such as the Indian Craton, the Himalayas, Eastern Tibet and subduction zones in the Pacific Ocean are not deprived of attention.

Of particular interest is the work devoted to the study of the deep distribution of electrical conductivity depending on the existing thermodynamic regimes at depth, as well as calculations of temperature and rheological parameters using geoelectro-magnetic and petrophysical data. A number of articles are devoted to theoretical works, computer simulations and examples of experimental observations. In addition, the Geological Institute KSC RAS has developed a set of programs for processing and interpreting the results of research in the field of electromagnetic sounding using controlled sources and sounding with natural fields. The developed algorithms were first tested in practice. Some projects related to future researches are presented also.

Presented for review the proceedings of the 2nd scientific-practical seminar "The Study of Continental Lithosphere Electrical Conductivity, Temperature and Rheology" edited by A. A. Zhamaletdinov and Yu. L. Rebetsky causes scientific interest. New researches are presents, new developments, both theoretical and methodical. The results obtained allow us to revise some existing ideas about the nature and nature of the thermodynamic characteristics of the lower horizons of the earth's crust.

Dr. of Sc. Valentina T. Filatova

Annotation

The main content of the articles' collection is devoted to the possibilities study for compilation of new models of the continental lithosphere structure by integrating the methods of geothermodynamics and deep geoelectrics. Considerable attention is paid to the study of nature of the deep geophysical boundaries using powerful controlled sources of the electromagnetic field. Of particular interest are researches related to the study of the transition boundary between the brittle and quasiplastic states of the matter of the earth's crust and the position of the creep area of the earth's crust. Geothermal and rheological studies in combination with the deep electromagnetic soundings are considered as a promising direction, which allows performing tectonophysical reconstruction of natural stresses in the lithosphere. The experimental studies' results and tectonophysical modeling are considered on examples of the Fennoscandinavian shield, the Indian Craton, the Himalayas, Eastern Tibet and the Eurasian continent as a whole. The collection is of interest to professional scientists involved in the study of Solid Earth Physics.

Contents

About the Editors

Abdullkhay A. Zhamaletdinov is DSc, Geological and Mineralogical Sciences (1991), professor of Murmansk Arctic State University, academician of the Russian Academy of Natural Sciences (2010). He was the principal investigator of more than 25 national and international research projects focusing on the study of the nature and structure of continental lithosphere electrical conductivity in complex with geodynamic, geothermal and rheological reconstructions with the use of super-deep drilling data. Most studies are conducted using powerful, controlled source electromagnetic soundings. Results of his research are reflected in 210 scientific articles, eight monographs and three patents.

Yury L. Rebetsky is DSc, Physical and Mathematical Sciences (PhD in Technical Sciences). He is a leading specialist of Russia in the study of natural stresses in the earth's crust and the author of the original method of tectonophysical inversion (reconstruction) of natural stresses from data on faults and cracks, as well as seismological data on the mechanisms of earthquake foci. They performed the reconstruction of the modern stress on seismically active regions of Eurasia. He is the head of the section "Tectonophysics" at the Department of Earth Sciences RAS. He is the one of the leading experts in the field of geomechanical and tectono-physical modeling of tectonic objects of the earth's crust. The results of his research are reflected in 82 scientific articles in leading Russian and foreign scientific journals, as well as four monographs.

Introduction

A. N. Vinogradov$^{(\boxtimes)}$

FIC KSC RAS, Krasnoyarsk, Russia

The history of deep electromagnetic soundings on the Kola Peninsula originates in the MHD-experiment "Khibiny", conducted under the scientific guidance of Academician Evgeny Pavlovich Velihov (1974–1990) (Velikhov 1989). The main result of the MHD experiment was the development of a "normal" model of the electrical conductivity of the lithosphere at a depth of 50–70 km. The result obtained allowed to conclude that the continental crust, with all its thickness, is characterized by high specific resistance and is "dry". This conclusion contrasted sharply with the results of magnetotelluric soundings (MTS), which develop the idea that the lower crust (deeper than 10–20 km) is characterized by high electrical conductivity. The increase in electrical conductivity obtained from the MTS results was explained by the presence of fluids as a result of the dehydration of rocks (Hyndman and Shearer 1989). After the MHD-experiment "Khibiny", a number of other deep-sounding experiments were carried out using powerful controlled sources. Among them are the experiments "Volga", "Zeus" and "FENICS". These studies confirmed the first results of the "Khibiny" experiment and at the same time made it possible to achieve significant progress in understanding the nature of the geophysical boundaries of the continental crust. The "SC-layer" of Semenov was chosen as an independent unit, representing the zone of distribution of sulphide and carbon-containing electronically-conductive rocks, limited by a layer of the upper crust 10–12 km thick. At the same time, an intermediate conductive fluid layer of dilatant-diffusion nature ("DD-layer") was found at a depth range from 2–3 to 7–10 km. The nature of the "DD layer" is explained by the penetration of meteoric water into the depths along the listric zones of fracturing and micro-cracks. The presence of free fluids at these depths is explained by the phenomenon of dilatancy—the irreversible opening of microcracks under the conditions of interaction of tangential (horizontal) and lithostatic (vertical) pressures. One of the most important results of the FENICS experiment was the establishment of a high horizontal uniformity of the middle layer of the bark along with its high specific resistance.

The precise study of the deep electrical conductivity allowed us to proceed to the study of the temperature parameters of the Earth's crust and to extrapolate the geothermal data of the Kola ultradeep well to the Moho border. Temperature studies, in turn, made it possible to proceed with the assessment of the rheological parameters of lithosphere, that is, the assessment of viscosity and plasticity of rocks at depth and, consequently, the assessment of the deep geodynamical regime of the Earth's interior (Zhamaletdinov 2011). Proceedings of this seminar are the first step to evaluate and common analysis of results of these studies.

© Springer Nature Switzerland AG 2019
A. A. Zhamaletdinov and Y. L. Rebetsky (Eds.): SPS 2018, SPEES, pp. 1–2, 2019.
https://doi.org/10.1007/978-3-030-35906-5_1

References

Velikhov, E.P. (resp. ed.): Geoelectric research with a powerful current source on the Baltic Shield. M. Science, p. 272 (1989)

Hyndman, R.D., Shearer, P.M.: Water in the lower continental crust: modeling magnetotelluric and seismic reflection results. Geophys. J. Int. **98**, 343–365 (1989)

Zhamaletdinov, A.A.: The new data on the structure of the continental crust based on the results of electromagnetic sounding with the use of powerful controlled sources. Dokl. Earth Sci. **438**, 798–802 (2011). (Part 2). ISSN 1028_334X

Theoretical Problems by Electrical Conductivity, Temperature and Rheology Researches

Modern Stress of the Crust of Eurasia

Yu. L. Rebetsky[(✉)]

Schmidt Institute of Physics of the Earth of the Russian Academy of Sciences,
Moscow, Russia
reb@ifz.ru

The method of cataclastic analysis of discontinuous dislocations created in the early 1990s (Rebetsky 1996, Rebetsky 1997) initially, as well as the methods of O. I. Gushchenko, J. Angelier, J. Gephard, had in his algorithm only the possibility of determining the shape and direction of the principal axes of the stress ellipsoid. But already at the beginning of the new century (Rebetsky 2003, Rebetsky 2009a, b) it was developed to obtain data on the ratio of the ball and deviator components of the stress tensor, and then to determine the absolute values of stresses, brittle strength of the cohesion and fluid pressure level. The determination of all these parameters required additional data in the form of generalizations of the Mohr diagram of the brittle fracture of faulting rock, seismological data on the value of the stress drop in the foci of the strongest regional earthquakes and data on lithostatic pressure at the depths of stress reconstruction.

1 Altai-Sayan (AS)

Based on the results of the reconstruction of natural stresses for crustal of AS, performed in collaboration with OA Kuchai (Rebetsky et al. 2013; Rebetsky et al. 2012), according to the data on the mechanisms of earthquake foci, it is established that the geochemical conditions of horizontal extension and shear in the horizontal plane (75% of the area of the troughs with data on the stress state most often correspond to the crust of large intra-mountain depressions, intermountain depressions and depressions). These conditions correspond to the sub-horizontal orientation of the axis of the principal deviatorial extension. The horizontal compression mode, which corresponds to the sub-vertical position of the axis of deviator extension, occurs much less frequently (the marginal parts of the Zaisan depression, the central segment of the Tuva depression, the Todzhi depression - 25% of the area of the depressions). We note that, due to the peculiarity of the seismic regime (Zhalkovsky et al. 1995), the amount of data on the mechanisms of earthquake foci for the regions of deflections of the crust is substantially smaller than for the regions of uplifts. This predetermined the large scale of stress averaging in the process of reconstruction and interpolation in the regionalization of the crust according to the types of geodynamic regime (Fig. 1).

At the same time, in the crust of uplifts - mountain ridges, anticline, massifs, etc. - the mode of horizontal compression and shear (65% of the area of uplifts), which corresponds to the sub-horizontal position of the axis of maximum compression, predominantly takes place. The situation of horizontal extension occurs much less

© Springer Nature Switzerland AG 2019
A. A. Zhamaletdinov and Y. L. Rebetsky (Eds.): SPS 2018, SPEES, pp. 5–12, 2019.
https://doi.org/10.1007/978-3-030-35906-5_2

frequently in the bark of the uplifts (the Sangilensk and Tannuol uplifts, the transition zone from the Mongolian Altai to the Mountainous Altai, the central part of the Western Sayan - 35% of the area of it's uplift).

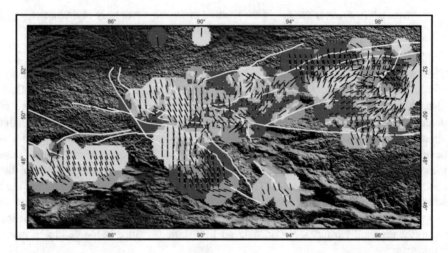

Fig. 1. Plunging compression axes and geodynamic types of stress state in the Altai and Sayan crust

It should be noted that the Tannuol Uplift divides the two largest intermountain depressions (East Tuva and Ubsu-Nurskaya). Probably, the seismic regime of these depressions makes a greater contribution to the determination of the mechanisms of foci of earthquakes in this region. From the same positions, one can also explain the extension of the regime of horizontal extension to the South Chuya Range near the Kuray and Chui Depressions of Gorny Altai. An even more important consequence of the analysis of the distribution of natural stresses is the established fact of the conjugate position of cortical sites with different geodynamic types of stress state.

2   Northern Tien Shan (STS)

The similar correlation between the morphology of the crustal roof and the geodynamic regime for STS is less pronounced. Here, the results of tectonophysical reconstruction, performed on the basis of the KNET network in conjunction with NA Sycheva (Rebetsky et al. 2012, Rebetsky et al. 2016a, b) show that in the crust at the eastern extremity of the Chui basin along its southern borders with the Kirghiz ridge, there is a region of horizontal extension. Suusamyr and Kochkor intra-mountain cavities correspond to the modes of horizontal shear or shear with stretching, for which the axis of the main deviatorial extension is sub-horizontal. The areas of horizontal extension also appear in the southern frame of the Kirghiz ridge near the northeastern border of the Suusamyr depression and near the eastern end of the Kochkor depression. In the crust of the mountain ranges themselves (the central part of the STS), the geodynamic regime, as a rule, is horizontal compression or shear with compression (Fig. 2).

It should be noted that in the western part of the region under investigation, there are regions of the horizontal extension for the section of the Kirghiz ridge. They are located on the northern and southern slopes of this ridge and are separated by a zone of horizontal shift. The appearance of these local zones of horizontal extension in the mountain ridge can be associated with the active fracture site at the present time.

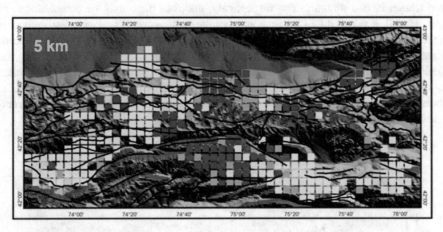

Fig. 2. Geodynamic types of stress in the upper part of the crust (0–10 km) in the northern part of the Central Tien Shan

3 High Asia (VA)

The established regularities in the distribution of the orientations of the main stresses in the core of the AS and STS were supplemented by one more, revealed by the results of reconstruction of the stresses in the cortex of Tibet, the Pamirs and the Himalayas - VA (Rebetsky and Alekseev 2014). As a result of the studies carried out in this work, it was established that the subregional orientation of the axis of the maximum deviatorial extension corresponds to the extensive sections of the crust of the central part of Tibet and the Pamirs, which are high-raised plateaus (4–5 km), which corresponds to the geodynamic type of stress state of horizontal extension and horizontal shift. In the environment of the crust of these plateaus, where the high mountain ranges of the Pamir and Tibet (Kunlun, Himalayas, etc.) are observed, there is a horizontal compression regime, which is also revealed for the AS and STS lift structures.

The presence of a horizontal orientation of the axes of the maximum deviatoric extension in the crust of uplifts in the form of the plateaus of Tibet and the Pamirs does not correspond to the above data for the AS and STS, where the subhorizontal orientation of the axes of maximum compression takes place in the crust of mountain rises. However, he does not refute them, but supplements them. It follows that mountain uplifts in the form of high mountain ranges and uplifts in the form of flat plateaus have a different type of geodynamic regime (Fig. 3).

Note that the bark of the Sangilensky uplift (highland), which is a relatively weakly elevated relief, also corresponds to the horizontal extension mode. Thus, an additional type of stressed state also found in the crust of the AS is found for the plateaus of Tibet and the Pamirs.

As mentioned above, there are cases when, in adjacent areas, the crust of the uplift and trough in one direction act, respectively, and, but there may be combinations of and, as well. It is noted that for such areas of the crust, a specific mosaic distribution of underthrusting shear stresses acting on horizontal areas is observed. The direction of these vectors should first of all suggest which parts of the uplift surrounding the depression actively interact with it. Here we should recall the "Karpinsky rule" determining that near the growing uplift, on the slopes of which erosion processes are intensively operating, there always exist deepening depressions that actively accumulate most of the sedimentary rocks.

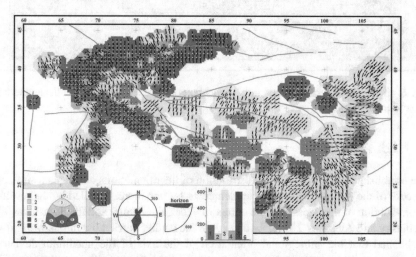

Fig. 3. Dipping compression axes and geodynamic types of stress state in the crust of high Asia

In conclusion of this part, we note that the results of tectonophysical reconstruction rely on seismological data on the mechanisms of earthquake foci, which in the investigated orogenes occur at depths from 3 to 20–25 km with the most representative location of 10–15 km. Therefore, the noted patterns of stress distribution in the crust of uplifts and deflections should also be attributed to the depths of the middle crust (10–20 km) and the adjacent portion of the upper crust (0–10 km) adjacent to it.

4 About the Dynamopar of the Stressed State of the Uplift and Depression of the Crust

According to the results of tectonophysical reconstruction of natural stresses, based on seismological data on the mechanisms of earthquake foci, it is established that: (1) the crust of the intra-continental mountain-fold orogenic STS and AS in the averaging scale comparable with the crustal thickness has a substantially heterogeneous stress-strain state; (2) in the crust of large areas of uplifts in the form of high mountain ranges in most cases (about 65% of their area), the stress state corresponds to the orientation of the compression axis in the sub-horizontal direction (geodynamic regimes of horizontal compression and shear), and in the crust of large sections of deflections in most cases about 75% of their area), the stress state corresponds to the orientation of the extension axis in the sub-horizontal direction (geodynamic regimes of horizontal extension and shear); (3) in the cortex of large areas and high mountain plateaus (Pamir, Tibet, Tuva upland), the orientation of the deviatorial extension axes is sub-horizontal; (4) the isotropic pressure value averaged over the area of the crust of the mountain-fold orogenic STS, AS, and VA are close to lithostatic pressure, which gives a level of additional (relative to the purely gravitational stress state) horizontal compressive stresses of about 1.3 kbar; (5) the spread in the results of calculation of the tectonic pressure is about 0.2–0.4 of the lithostatic pressure, which leads to the appearance of regions of a high level of deviatoric stresses both in zones of increased tectonic pressure and where it decreases.

5 Magnitudes of Tectonic Stresses

In the tectonophysical studies of seismically active regions conducted at the beginning of the zero years using the results of the reconstruction of natural stresses (Rebetsky 2003), it was shown that a sufficiently mosaic pattern of stress distribution corresponds to seismogenic fault zones. Such a mosaic in stresses is, of course, connected with the peculiarity of the structural-material state of different fault sites (Rebetsky 2005). Already the first results made it possible to draw attention to the fact that critical stress states in different parts of the Mohr diagram must have different mechanisms for dissipating mechanical energy: (1) a strong earthquake - a large-scale brittle fracture of the earth's crust; (2) the set of medium-strong and weak earthquakes - cataclastic (pseudo-plastic - mech.) Or quasibrate flow along the extended zone of the crustal fault; (3) quasistatic creep along the fault - quasi-viscous or quasiplastic (due to microcracks at the level of grains) current (Rebetsky 2003, Rebetsky 2005). In the conducted researches it was established:

(1) Stresses in magnitude are distributed in the earth's crust extremely unevenly, their difference in one seismic focal area can be as high as 1–1.5;
(2) There is a certain range of relationships between the level of effective pressure and the maximum shear stress of 0.5–2, related to the requirement of the Coulomb-Mohr criterion for brittle fracture;

(3) The level of fluid pressure in the uppermost horizons of the crystalline crust (1–5 km) is most differentiated (1–2 from the hydrostatic level), and in the deeper horizons approaches lithostatics;

(4) The strength of the adhesion of real fractured arrays, corresponding to the averaging scale of the calculated stresses in the first tens of kilometers, is much lower than that of whole samples of crystalline rocks measuring 5–10 cm and varies in the range of 1–50 bar;

(5) The brittle strength of arrays of intra-continental orogens is higher than that in subduction zones of lithospheric plates 3–5 times, which determines a higher level of deviatorial stresses and effective pressure in the continental crust (Fig. 4);

(6) The level of the maximum shear stresses in subduction zones generating the strongest earthquakes ranges from 3–5 bar to 50–100 bar;

(7) The relationship between the tectonic pressure and the maximum shear stress level has the greatest dispersion in the cortex of intra-continental orogens and is close to linear in the subduction zones of lithospheric plates.

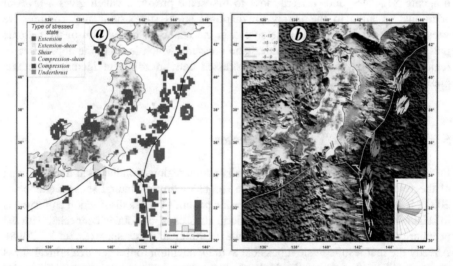

Fig. 4. Geodynamic types of the stressed state in the subduction zone of Japan (a) and the orientation of the axes and the relative values of the maximum horizontal compression (b)

In the results of the reconstruction of the stresses of different seismically active regions, important regularities were also obtained that carry information about the mechanisms of their deformation (Rebetsky et al. 2013, Rebetsky and Polets 2014, Rebetsky 2015, Rebetsky et al. 2012, Rebetsky et al. 2016a, b). In the annex to the problem of seismic hazard, it was found that relatively strong regional earthquakes rarely fall into the zones of an increased level of maximum tangential stresses. Their epicenters, as a rule, are located in the zones of the middle level of this stress or in zones of its elevated gradient. This empirically observed fact coincides with the replica of Rice from his monograph (Rice 1982), in which it was suggested that "strong earthquakes should not fall into the region of an increased stress level, here, on the planes of rupture, there is a high level of friction, overcoming which will take most of the released energy."

6 Conclusions

The results of the conducted studies showed that the state of the crust of depressions and uplifts as separate ridges and their groups in the areas of intra-continental orogenes is antipodal. Various types of studies, a review of which was made in the article, showed that in the crust of actively growing uplifts and caving depressions, a different geodynamic type of stress state. In the application to orogenes, this means that in the crust of ridges the maximum compression is predominantly sub-horizontal, and in the crust of intermountain or large intra-mountain depressions, the axis of the minimum compressive stress (maximum deviator extension) is predominantly oriented in the sub-horizontal direction. The level of isotropic compression in the crust of the developing uplift is higher than in the neighboring sections of the crust.

The performed estimates of the magnitude of the additional horizontal stresses necessary to change the geodynamic type of the stress state of horizontal extension to horizontal compression give a value of about 5.4 kbar. For the velocities of lateral deformations, measured today by GPS-geodesy, it will take up to 10 million years for lateral deformations to create a similar level of stress. At stresses corresponding to the yield strength of rocks for the geodynamic type of stress state of horizontal compression, 20–25 million years are needed to create a thickened crust at 55–60 km. Thus, in the orogens, areas of the cortex with two essentially different states often coexist, in one of which the magnitude of the isotropic pression (the crust of ridges and uplifts) is substantially higher than the level of isotropic pression of the neighboring cavity. A similar situation for the performance of mechanical equilibrium requires the appearance near the Moho boundary on horizontal areas of shearing tangential stresses, oriented in different directions from the depression (for sites with normals directed to the center of the earth). Since in the orogenes the basins and mountain uplifts are sufficiently mosaically distributed, this should also lead to a mosaic in the orientation of such tangential stresses. The results of tectonophysical stress reconstructions (Rebetsky et al. 2013, Rebetsky et al. 2012, Rebetsky and Alekseev 2014) showed that the greatest variability and mosaic in the distribution of the orientation of the underthrusting shear stresses is observed exactly where there are large depressions.

References

Rebetsky, Yu.L., Alekseev, R.S.: The field of recent tectonic stresses in central and South-Eastern Asia. Geodyn. Tectonophys. 5(1), 257–290 (2014). https://doi.org/10.5800/GT-2014-5-1-0127

Rebetsky, Yu.L.: Tectonic stress, metamorphism, and earthquake source model. Dokl. Earth Sci. 400(1), 127–131 (2005)

Rebetsky, Yu.L., Sycheva, N.A., Sychev, V.N., Kuzikov, S.I., Marinin, A.V.: The stress state of the northern Tien Shan crust based on the KNET seismic network data. Russ. Geol. Geophys. **57**, 387–408 (2016a). https://doi.org/10.1016/j.rgg.2016.03.003

Rebetsky, Yu.L.: Estimation of stress values in the method of cataclastic analysis of shear fracture. Dokl. Earth Sci. **428**(7), 1202–1207 (2009a)

Rebetsky, Yu.L.: Stress-monitoring: Issues of reconstruction methods of tectonic stresses and seismotectonic deformations. J. Earthq. Predict. Res. **5**(4), 557–573 (1996). Beijing, China

Rebetsky, Yu.L.: On the specific state of crustal stresses in intracontinental orogens. Geodyn. Tectonophys. **6**(4), 437–466 (2015). https://doi.org/10.5800/GT-2015-6-4-0189

Rebetsky Yu.L.: Reconstruction of tectonic stresses and seismotectonic strain: methodical fundamentals, current stress field of Southeastern Asia and Oceania. Translation (Doklady) of the Russian Academy of Science, vol. 354, issue 4. pp. 560–563 (1997)

Rebetsky, Yu.L.: Stress state of the earth's crust of the Kurils Islands and Kamchatka before the Simushir earthquake. Russ. J. Pac. Geol. **3**(5), 477–490 (2009b)

Rebetsky, Yu.L.: Stress-strain state and mechanical properties of natural massifs according to the data on the mechanisms of earthquake foci and the structural-kinematic characteristics of cracks. Avtoref. dis. doc. phys. m. sciences, p. 56. Izd. OIFZ, Moscow (2003)

Rebetsky, Yu.L., Kuchai, O.A., Marinin, A.V.: Stress state and deformation of the earth's crust in the Altai-Sayan mountain region. Russ. Geol. Geophys. **54**(2), 206–222 (2013). https://doi.org/10.1016/j.rgg.2013.01.011

Rebetsky, Yu.L., Polets, A.Yu.: The state of stresses of the lithosphere in Japan before the catastrophic Tohoku earthquake of 11 March 2011. Geodyn. Tectonophys. **5**(1), 469–506 (2014)

Rebetsky, Yu.L., Polets, A.Yu., Zlobin, T.K.: The state of stress in the earth's crust along the northwestern flank of the pacific seismic focal zone before the Tohoku earthquake of 11 March 2011. Tectonophysics **685**, 60–76 (2016b). https://doi.org/10.1016/j.tecto.2016.07.016

Rebetsky, Yu.L., Sycheva, N.A., Kuchay, O.A., Tatevossian, R.E.: Development of inversion methods on fault slip data. Stress state in orogenes of the central Asia. Tectonophysics **581**, 114–131 (2012). https://doi.org/10.1016/j.tecto.2012.09.027

Rice, J.: The mechanics of earthquake rupture. In: Dziewonski, A., Boschi, E. (eds.) Physics of the Earth's Interior, pp. 555–649. Elsevier, Amsterdam (1982)

Zhalkovsky, N.D., Kuchai, O.A., Muchnaya, V.I.: Seismicity and some characteristics of the stress state of the earth's crust of the Altai-Sayan region. Geol. Geophys. **36**(10), 20–30 (1995). Russian

On the Nature of the Brittle-Ductile Transition Zone in the Earth's Crust (Review)

A. A. Zhamaletdinov[1,2]([⊠])

[1] St. Petersburg Brunch of IZMIRAN, St. Petersburg, Russia
abd.zham@mail.ru
[2] Geological Institute KSC RAS, Apatity, Russia

Abstract. A short review of modern rheological and geophysical researches on the problem of the brittle-ductile transition zone existence in the Earth's crust is undertaken in the introduction of the article. The first part of presentation is devoted to summarization of electromagnetic sounding results, performed with the use of powerful controlled sources. In the second part the geodynamical interpretation is given for obtained results. It is proposed to divide the continental lithosphere into two parts - upper and lower. The upper part is brittle and comparatively low resistive (10^4 Ω). It has the thickness of about 10–15 km and is most actively involved in geological processes. Its principal peculiarities are - the sharp horizontal heterogeneity and a broad range variations of rocks resistivity (from 1 to 10^5 Ω·m), the contrasting character of geological structures, a wide distribution of fault zones and fractures, high brittleness, the presence of fluids that penetrate to the depth from the day surface (DD layer), and wide distribution of electronic-conducting structures ("SC layer" of Semenov) in the composition of volcanogenic-sedimentary sequences. The lower part of the Earth crust (from 10–15 to 35–45 km) is highly resistive (10^5–10^7 Ω·m) and horizontally homogeneous. The porosity and content of free fluids at these depths reduces sharply. Rocks are in a ductile or semi-ductile state. All these facts signify that the electric conductivity at depths of more than 10–15 km is determined more by physical parameters (pressure, temperature, viscosity) then by geological composition.

Keywords: Brittle · Ductile · Geoelectrics · Fennoscandian shield · Reology

1 Introduction

Academician Sadovsky (1945) was the first who developed the idea of a brittle state of the earth's crust in his researches devoted to justifying the law of similarity of the earth's crust elements observed in large explosions and the problem of "lumpiness" in the earth's crust. He considered "lumpiness" as a multidimensional separation of the structural links of the earth's crust as a type of self-similar fractal structures. This idea found its continuation in subsequent works devoted to the analysis of the stress-strain state of the environment (Gzovsky 1975, Nikolayevsky 1996). A large number of works have been devoted to rheological parameters study of the medium on the basis of complexation of a wide range of geophysical characteristics of the medium (elasticity, density, magnetic properties, geothermy, heat generation, etc.) (Ranalli 2000, Glasnev

A. A. Zhamaletdinov and Y. L. Rebetsky (Eds.): SPS 2018, SPEES, pp. 13–21, 2019.
https://doi.org/10.1007/978-3-030-35906-5_3

2003, Moisio and Kaikkonen 2006). In general, there composed an idea of a two-layered structure of the earth's crust consisting of an upper, brittle 'layer and a lower ductile layer. The depth to the boundary between the brittle and plastic states of the earth's crust is the most important parameter and, at the same time, it is the most difficult for the quantitative estimation, because it depends on many a priori factors obtained from laboratory data. A two-layered model of earth's crust structure finds confirmation in the results of seismology. It is believed that the upper, seismically active layer where short-focus earthquakes are generated is more brittle. On the territory of the Fennoscandian shield it is limited to the upper 10–12 km of the thickness of the earth's crust (Korhonen and Porkka 1981; Sharov 2017). Kissin (1996) first suggested that at a depth of 10–12 km there should be a boundary of "impenetrability" between mantle fluids penetrating the earth's crust from down to up due to dehydration of deep mafic rocks and meteoric fluids penetrating into the earth crust from the top to the down along the system of subvertical and inclined (listric) fault zones.

2 Geoelectromagnetic Researches

Certain expectations in assessing the nature of geophysical boundaries in the crystalline earth's crust were associated with geoelectromagnetic methods, in particular, with magnetotelluric sounding (MTS). It was supposed "a priori" that seismic boundaries should be accompanied by abrupt increases in electrical conductivity in the form of intermediate conducting layers, the nature of which can be related to temperature gradients, changes in the chemical (petrographic) composition of rocks or changes in fluid saturation of the lithosphere (Vanyan and Gliko 2002). However, no reliable solutions have been obtained to date due to the multifactor nature of the interpretation of the results obtained, in connection with the ambiguity of the solutions of the inverse problem of electromagnetic soundings and several restrictions from petrology (Yardley and Valley 1997) and tectonophysics (Nesbitt 1993).

To date, two alternative views have formed on the electrical conductivity of the continental crust. They are shown in Fig. 1 in the shape of two schematic models.

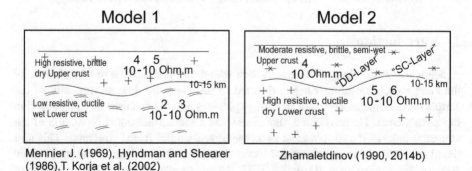

Mennier J. (1969), Hyndman and Shearer (1986), T. Korja et al. (2002)   Zhamaletdinov (1990, 2014b)

Fig. 1. Two alternative models of the Fennoscandian lithosphere geoelectrical structure

Model 1 is developed mainly on the base of soundings in the field of natural sources (AMT-MTS). The upper part of the earth's crust in the model 1 is poorly conductive, brittle and dry, and the lower part, deeper than 10–20 km, is conductive, plastic and contains free fluids that appear at depth due to dehydration of the upper mantle (Hyndman and Shearer 1989, Kissin 1996). Model 2 is developed mainly on the base of soundings, performed in the field of powerful controlled sources (Zhamaletdinov 1990, 2014b). The upper crust of 10–15 km thickness has a moderate average resistivity of about 10^4 Ω·m. Meteoric waters penetrate into it and form an intermediate conductive layer of dilatancy-diffusive nature ("DD-layer"), which enjoys regional distribution (Zhamaletdinov et al. 2017). The longitudinal conductivity of the "DD layer" is around of units and tenth of Siemens. Against the backdrop of the "DD layer", sulfide-carbon electronic-conducting rocks of the "SC-layer" of Semenov are widely spread in the upper stratum of the earth's crust. Their longitudinal conductivity reaches thousands of Siemens and linear dimensions reach hundreds and sometimes thousand of kilometers (Zhamaletdinov 2014a). Nevertheless, the "SC layer" does not have a significant effect on the integrated longitudinal conductivity of the earth's crust, since separated electronic conductors, as a rule, are not electrically connected between themselves and do not represent a single uninterrupted conducting layer.

In the present paper, the existence of a boundary of another origin is supposed in the earth's crust. It exists in the form of a poorly conductive basement, as a boundary of "impermeability" for galvanic currents at a depth of 10–15 km It is assumed that this boundary marks a transition zone between the upper, brittle zone of the earth's crust and the lower, ductile zone. It is idefined conditionally as the Conrad zone sporadically detected by seismic data (Zhamaletdinov 2014a).

From induction electromagnetic soundings theory (Vanyan 1997) It is wellknown that high-resistive layers, if their thickness is less than the length of electromagnetic wave in limits of quasi-stationary wave zone, fall into the region of "transparency". Their detection in this case, in particular, in magnetotelluric sounding (MTS), becomes problematic. As an illustration, Fig. 2 shows an example of calculating the direct MTS problem for four models of the geoelectric section of type "K" (M1-M4) with varying parameters of an intermediate poorly conducting layer with a thickness of 20 km (Fig. 2b).

Model M2 is a "normal" geoelectric section of the Fennoscandian shield, obtained from the results of the "FENICS" experiment (Zhamaletdinov et al. 2011). The M1 model has an abnormally low (by 0.5 order) resistivity compare to M2 model. The sections on the M3 and M4 models, on the contrary, have an anomalously high resistivity (10^6 and 10^7 Ω·m) in comparison with the M2 model. The amplitude and phase curves of the MTS for the M3 and M4 models are shown in Fig. 2a. They differ slightly from the M2 model and practically do not differ among themselves. This means that the resistivity of the intermediate layer can be arbitrarily large, but this will not affect the character of the MTS curves, either in amplitude or in phase. The same is true for frequency soundings with controlled sources in a quasi-stationary wave zone.

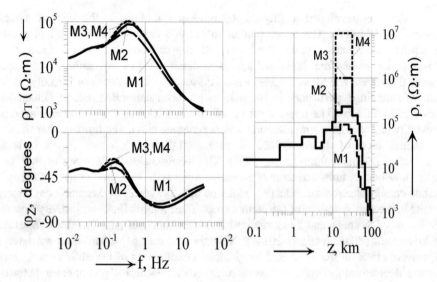

Fig. 2. An example of curves calculation for apparent resistivity and phase of impedance by magnetotelluric sounding (MTS) (a) for 4 models of electrical section with an intermediate high resistive layer (b)

The curves of apparent resistivity on DC sounding (Fig. 3a) are significantly more informative compared to the MTS curves. You can see that they have different character for different models of section. From their character and amplitude two parameters of vertical electrical section can be determined at least. The first parameter is the depth of the roof of the poorly conducting layer h. The value h is approximately equal to the distance between the source and the receiver, on which a sharp increase in the gradient of the curve occurs (Fig. 3a).

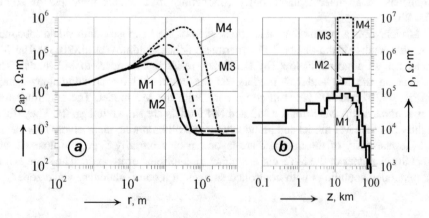

Fig. 3. An example of calculation of the apparent resistivity curves of the deep DC sounding (a) for 4 models of an electrical section with an intermediate high resistive layer (b).

The transversal resistance T of the poorly conducting layer of the earth's crust is the second parameter. The value of T is determined by the Eq. (1),

$$T = \rho_a^{max} \cdot r_{oo'}^{max} \tag{1}$$

where ρ_a^{max} is the maximum value of the apparent resistivity at the separation $r_{oo'}^{max}$, after which the apparent resistivity curve acquires a descending form, indicating the initiation of DC leakage through the poorly conductive layer into the lower conducting base. In particular, according to the "normal" curve (model M2 in Fig. 3a), it can be determined that the minimum value of the transversal resistance T of the Baltic shield's lithosphere is about of $7 \cdot 10^9$ $\Omega \cdot m^2$.

The advantages of DC sensing mentioned above exist only theoretically. In practice, their use faces various technical and methodological difficulties, among which the main is the influence of the horizontal heterogeneity of the environment, which inevitably arises when the distance between the source and receiver in different geological conditions increases. Therefore, until now, the problem of determining the location and depth parameters of the poorly conducting "impenetrability" boundary, with which we associate the boundary between the fragile and quasiplastic states of the earth's crust, has not been solved. The existence of this boundary was previously assumed at the semi-empirical level of knowledge on indirect grounds.

The first experimental evidence that in the poorly conductive crystalline rocks of the Baltic shield even more poorly conductive, relatively "impenetrable" base exists at depths greater than 10 km, has been obtained by means of sounding with the use of impulsive 80 MW MHD-generator (Zhamaletdinov 1990). Two indicators served as the basis for such a conclusion. The first indicator is depth of subvertical electronically conductive zones penetration to the earth crust. These zones are associated with current conductive channels. Their penetration depth does not exceed 10 km by different estimates (Zhamaletdinov 1990, Zhdanov and Frenkel 1983, Kirillov and Osipenko 1984). The second indicator based on theoretical estimates that show a subhorizontal (in the cylindrical coordinate system) spreading of the galvanic currents of the MHD generator in the upper film of the Earth crust of about 10 km thickness.

Subsequent experiments, first of all, the FENICS frequency soundings with the use of two mutually orthogonal industrial transmission lines (LEP) of 10^9 and 120 km long and with a generator of 200 kW (Zhamaletdinov et al. 2011) and frequency sounding in Central Finland and in the Kola Peninsula with the car-generator of 29 kW power (Zhamaletdinov et al. 2002 and 2017) made it possible to obtain additional experimental material to substantiate the hypothesis of a two-layered structure of the earth's crust into a brittle and quasi-plastic region with separating them "the boundary of impenetrability".

3   Geodynamical Interpretation

Geodynamical interpretation is presented on the Fig. 4 in the generalized form. On the left panel (Fig. 4a) the so-called "normal" electric section of the Baltic Shield lithosphere is given. The section is obtained in conditions of absence of the influence from electronically conductive rocks.

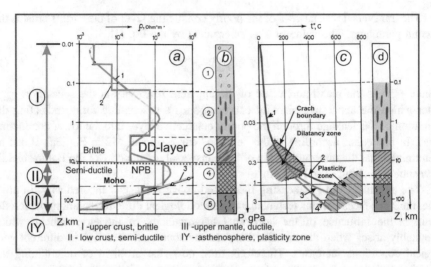

Fig. 4. Geodynamic model of the structure of the continental lithosphere according to the results of electromagnetic soundings with the use of powerful controlled sources. Explanations to the figure are given in the text

The Fig. 4a is represented in the form of a gradient model (curve 1) and its layered approximation (curve 2). The resulting section belongs to the KHK-type model with three conductive layers. Its interpretation is given on the Fig. 4b in the form of a structural-geodynamic column. The upper part of the section (layer 1) consists of conductive sedimentary deposits (moraine) and a water filled part of the crystalline basement's roof. The high resistive layer 2 consists of sub vertical faults and cracks filled up by the water solutions (fluids). The average thickness of the layer 2 is esti-mated by 2–3 km. In the depth interval from 2 to 10 km the intermediate conductive layer (3) is situated. Its resistivity decrease from about 10^5 Ω·m to about 10^4 O·m and less. The cause of this decrease in resistivity is related to fluids of meteoric origin. They penetrate from day-time surface into depths along cracks and faults flattening with depth. As it was mentioned above, this layer has a dilatancy-diffusive nature and it is identified as "DD layer". Deeper than 10 km resistivity of rocks increases again up to 10^5-10^6 Ω·m. Here locates the boundary of impermeability (BIP), dividing the earth crust into upper (brittle) and lower (semi ductile, ductile) parts.

The high resistive part of the lithosphere locates in the depth range from 10–20 to 60–80 km. The average value of the transversal resistivity T is of about 10^{10} Ω·m². The exponential decrease of resistivity occurs below 60–80 km under the influence of temperature increasing with depth. At the depth range of 200–300 km the resistivity of lithosphere fell down up to 100 Ω·m, pointing out on the possible existence of the asthenosphere.

On the right panel (Fig. 4c) the geodynamic diagram is given after (Nikolayevsky 1996). Here the temperature profiles are given: curve 1 is measured in the Kola super deep hole, (Popov et al. 1999), curves 2–4 – results of its extrapolation: curve 2 (Kremenetsky et al. 1998), curve 3 (Zhamaletdinov 1982), curve 4 (Valle 1951). The dilatancy zone occupies the depth range 5–15 km on the theoretical diagram (Fig. 4c, d) and 2–10 km on geoelectrical panel (Fig. 4a, b). Plasticity zone occupies 40 km and more on the right panel (Fig. 4c, d) and 80 km and more on the left panel (Fig. 4a, b). Clarification of the reasons for these discrepancies may be the subject of further research.

Results of experimental study of brittle-ductile transition zone, connected with the boundary of impermeability, are presented in this collection in the article "Zhamaletdinov A.A., Shevtsov A.N., Skorokhodov A.A., Kolobov V.V., Ivonin V.V. Experimental study of impermeability boundary in the Earth crust".

4   Conclusion

The summarized analysis of presented above results of electromagnetic soundings has allowed us to divide the continental lithosphere into two parts- upper and lower. The upper part – brittle and more conductive –represents the Earth crust namely. It has the thickness of about 10–15 km and is most actively involved in geological processes. Its principal peculiarities are - the sharp horizontal heterogeneity and a broad range of specific electrical resistivity variations of rocks (from 1 to 10^5 $\Omega \cdot$m), the contrasting character of the structure of geological units, a common distribution of fault structures, higher brittleness, the presence of fluids that drain the super crustal from the day surface owing to supply of meteor waters to depth (DD layer), and common distribution of electronic-conducting structures ("SC layer" of Semenov) in the composition of volcanogenic-sedimentary sequences. The lower part (from 10–15 to 35–45 km) is high resistive (10^5–10^6 $\Omega \cdot$m) and horizontally homogeneous. The rocks at these depths form a ductile zone; the porosity and content of free fluids in them reduces sharply. All these facts signify that the electric conductivity at depths of more than 10–15 km is determined mostly by the planetary physical-chemical parameters (pressure, temperature, viscosity), the phase transitions of the substance, and the geodynamic peculiarities of evolution of different segments of the Earth, rather than by the geological processes observed near the day surface.

Acknowledgments. This work was done with financial support of RFBR grant 18-05-00528 (theory) and partly with the support of the State Mission GI KSC RAS, the number of research 0226-2019-0052 (interpretation). Author is deeply grateful to the lead programmer T.G. Korotkova for her help in results processing.

References

Glaznev, V.N.: Complex geophysical models of a lithosphere of the Fennoscandian. Apatity. "K&M". 252 p. (2003)

Gzovskiy M.V.: Fundamentals of tectonophysics. Science 535 (1975)

Hyndman, R.D., Shearer, P.M.: Water in the lower continental crust: modeling magnetotelluric and seismic reflection results. Geophys. J. Int. **98**, 343–365 (1989)

Kirillov, S.K., Osipenko, L.G.: Study of the Imandra-Varzuga conductive zone (Kola Peninsula) with the use of MHD generator. Crustal anomalies of electrical conductivity. L. Nauka, pp. 79–86 (1984). (in Russian)

Kissin, I.G.: Fluid saturation of the earth's crust, electrical conductivity, seismicity. Izv. AN SSSR, ser. Physics of the Earth. **4**. 30–40 (1996)

Korhonen, H., Porkka, M.T.: The structure of the baltic shield region on the basis of DSS and earthquake data. Pure. appl. Geophys. **119**(6), 1093–1099 (1981)

Korja, T., Engels, M., Zhamaletdinov, A.A., Kovtun, A.A., Palshin, N.A., Smirnov, M.Yu., Tokarev, A.D., Asming, V.E., Vanyan, L.L., Vardaniants, I.L., BEAR WG.: Crustal conductivity in Fennoscandia – a compilation of a database on crustal conductivity in Fennoscandian shield. Earth Planets Space. **54**, 535–558 (2002)

Kremenetsky, A.A., Ikorsky, S.V., Kamensky, I.L., Sazonov, A.M.: Geochemistry of the deep zones of the Precambrian crust. Kola superdeep. Scientific results and experience of research. (edited by V.P. Orlov and N.P. Laverov). M.M. 1998. Ed. Min Nat Res of RF, RAS and RANS. 255 p. (1998)

Mennier, J.: Sontage electric de la crounte ferrestre utilisant les courants de retour industriels. Compte Rendu Acad. Sci., № 6, v.268, ser. Act. B. pp. 514–516 (1969)

Moisio, K., Kaikkonen, P.: Three-dimensional numerical thermal and rheological modelling in the central Fennoscandian Shield. J. Geodyn. **42**(4–5), 95–114 (2006)

Nesbitt, B.E.: Electrical resistivities of crustal fluids. J. Geophys. Res. **98**(B3), 4301–4310 (1993)

Nikolayevsky, V.N.: Geomechanics and Fluid Dynamics. M. Nedra. 6 p. (1996)

Popov, Y., Pevzner, S.L., Pimenov, V.P., Pevzner, L.A.: Geothermal characteristics of the Kola superdeep well. DAN **369**(6), 823–826 (1999)

Ranalli, G.: Rheology of the crust and its role in tectonic reactivation. J. Geodyn. **30**, 3–15 (2000)

Sadovsky, M.A.: Experimental studies of the mechanical effect of an explosion shock wave. In: Proceedings of the Seismological Institute of the Academy of Sciences of the USSR, M.-L., no. 116 (1945)

Sharov, N.V.:. Lithosphere of Northern Europe by seismic data. Ed. Karelian Sci Centre of RAS. Petrozavodsk. 170 p. (2017)

Valle, P.E.: Sull'aumento ditempera nel mantello della terra per compressions adiabatica. Ann. Geofis. **5**(4), 475–478 (1951)

Vanyan, L.L.: Electromagnetic soundings. Moscow. "Nauchny Mir". 218 p. (1997)

Vanyan, L.L., Gliko, A.O.: Seismic and electromagnetic evidence of dehydration as a free water source in the reactivated crust. Geoph. J. Int. **137**(1), 159–162 (2002)

Yardley, B.W.D., Valley, J.W.: The petrologic case for a dry lower crust. J. Geophys. Res. B **6**, 12,173–12,185 (1997)

Zhamaletdinov, A.A., Shevtsov, A.N., Korotkova, T.G., et al.: Deep electromagnetic sounding of the lithosphere in the eastern baltic (Fennoscandian) shield with high_power controlled sources and industrial power transmission lines (FENICS experiment). Izvestiya. Phys. Solid Earth. **47**(1), 2–22 (2011)

Zhamaletdinov, A.A.:. The largest in the world anomalies of electrical conductivity and their nature - a review. Glob. J. Earth Sci. Eng. **1**, 84–96 (2014a)

Zhamaletdinov, A.A.: Normal electrical section of the crystalline basement and its geothermal interpretation from the data of MHD sounding on the Kola Peninsula. in the book: Deep EM soundings with the use of pulsed MHD generators. Apatity. 1982. The Kola. Phil. AN SSSR, pp. 35–46 (1982)

Zhamaletdinov, A.A.: Model of electrical conductivity of lithosphere by results of studies with controlled sources (Baltic shield, Russian plateform). Leningrad. "Nauka", 159 p. (1990)

Zhamaletdinov, A.A.: The nature of the conrad discontinuity with respect to the results of kola superdeep well drilling and the data of a deep geoelectrical survey. Doklady Earth Sci. Part 1 **455**, 350–354 (2014b)

Zhamaletdinov, A.A., Shevtsov, A.N., Tokarev, A.D., Korja, T.: Electromagnetic frequency sounding of the earth crust beneath the central finland granitoid complex. Izvestiya, Phys. Solid Earth, **38**(11), 954–967 (2002)

Zhamaletdinova, A.A., Velikhov, E.P., Shevtsov, A.N., Kolobov, V.V., Kolesnikov, V.E., Skorokhodov, A.A., Korotkova, T.G., Ivonin, V.V., Ryazantsev, P.A., Biruly, M.A.: . The Kovdor-2015 experiment: study of the parameters of a conductive layer of dilatancy–diffusion nature (DD Layer) in the archaean crystalline basement of the baltic shield. Doklady Earth Sci. Part 2 **474**, 641–645 (2017)

Zhdanov, M.S., Frenkel, M.A.: Migration of electromagnetic fields in solving inverse problems of geoelectrics. DAN USSR. **271**(3), 589–594 (1983)

Modeling of the Stress-Strain State of the Medium with Various Geomechanical and Rheological Parameters in the Annex to the Problems of Regional Geodynamics

D. S. Myagkov[⊠]

Schmidt Institute of Physics of the Earth of the Russian Academy of Sciences,
Moscow, Russia
dsm@ifz.ru

Abstract. In this paper, we present the results of numerical modeling of the formation of a stress-strain state, obtained both for regional geodynamic and for local geomechanical models. The first group includes the model of the formation of a stress-strain state in the subduction zone of the Tohoku area. The stress-strain state study in the subduction zone of the Tohoku area was carried out by numerical modeling. An explicit finite-difference scheme developed by Wilkins for the study of elasto-plastic bodies was applied and improved by Stefanov for application in geomechanics. The aim is to determine the geodynamic factors responsible for the formation of the stress state of the lithosphere subduction zones Tohoku region (northern Honshu) at a stage immediately preceding catastrophic earthquake Tohoku (2011). According to tectonic reconstructions, in this area there is the following pattern: at the lithosphere Japan microcontinent and oceanic lithosphere subducting beneath Honshu (to the west of the trenchcenter) observed the situation of lateral compression, whereas to the east of the trench is observed the opposite situation - the lateral stretching. Such regularity requires a geodynamic explanation, the search for which was done within the framework of the current work by the method of numerical simulation. The models of stress-strain state of the region, which is formed by the action of the small-scale convection asthenosphere, the lateral pressure from the Pacific plate. Exogenous processes (denudation and accumulation of the geomaterial) are studied separately.

The current study presents the results of numerical simulation of the formation of a stress-strain state in the Tohoku region of the Japanese subduction zone (North Honshu). This area is known for the catastrophic Tohoku earthquake that occurred in 2011, which resulted not only in the enormous economic damage inflicted on Japan and neighboring countries (including Russia), but also on a noticeable restructuring of the tense state in this sector of the subduction zone. The presence of a dense network of seismological stations in the region made it possible to obtain [1] a detailed distribution of natural stresses by methods of tectonophysics, which was to a large extent facilitated by the presence of seismic stations in the oceanic zone. The purpose of this work was the creation of various geodynamic models for the formation of the stress-strain state of

© Springer Nature Switzerland AG 2019
A. A. Zhamaletdinov and Y. L. Rebetsky (Eds.): SPS 2018, SPEES, pp. 22–26, 2019.
https://doi.org/10.1007/978-3-030-35906-5_4

the region, in order to determine the possibility of creating a tense state observable by natural data, presumably responsible for the formation of the Japanese zone of sub-duction of geodynamic processes. Modeling (the model scheme in Fig. 1) was per-formed on relatively generalized models, which allows using the results obtained also to study the origin of other subduction zones with a similar stress distribution structure, such as the Andean zone.

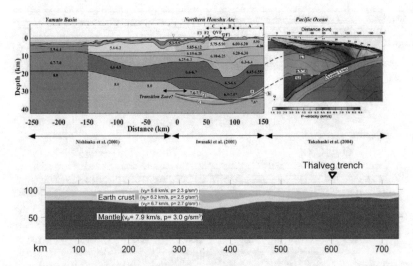

Fig. 1. The generalized geophysical profile and the model scheme constructed with it with the indicated values of the density and velocity of P-waves for the basic structural complexes of the model

In this paper, numerical models for the formation of excessive stresses of horizontal compression are also presented, the explanation of which is an important tectonic problem in the Earth's crust. Its discussion has been conducted practically since the beginning of the second half of the 20th century and has applied value, because these stresses are often one of the most serious problems in the development of ore deposits [2], including in the Russian territory (the Kola Region). The mechanisms of generation of an elevated level of horizontal stress were also considered, as a result of a vertical uplift of the crustal material [3]. Such uplift in regions with active demolition of sedimentary material can, first of all, occur due to denudation processes.

In the current study, two numerical elastic-plastic models are presented for the formation of additional "denudation" horizontal stresses. The aim of the work was to quantitatively study this effect for models, a significant part of which is in a super-critical state and show how it is formed, including for various strength properties of the medium and for various types of given plasticity law. The first model (Fig. 1) is simpler: a horizontally layered medium for which only the denudation process is specified, the location of the sedimentation area is assumed outside the model. The second model (Fig. 2) is laterally heterogeneous, the subduction zone of the Northern

Honshu region is considered. If for the first, "one-dimensional" model the task was set to describe the mechanism being studied directly (including for comparison with analytical models), in the second one an attempt was made to show its effect on the stress state of a real geological structure.

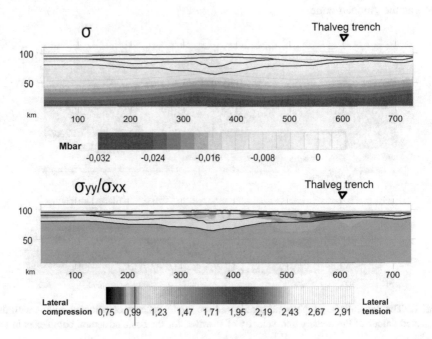

Fig. 2. Results of modeling for the effect of small-scale thermogravitational asthenospheric convection (with a peak amplitude of 10 km) after erosion. The field of medium stresses (pressure with the opposite sign) is shown from above in the model; bottom - the ratio of vertical and horizontal stresses

The simulation was carried out according to a technique developed by Wilkins for the study of elastoplastic bodies [4] and improved Stefanov [5] for application in geomechanics. The peculiarities of this approach are the use of an explicit finite-difference scheme and the recording of the equations of motion in a dynamic form, with further adaptation for the description of quasistatic processes. The bodies of the model are considered as elastic-plastic, rather than viscous, for the plasticity that is more appropriate to the geodynamic processes, the non-associated elastic-plastic law of the Drucker-Prager-Nikolaevsky flow is used where, in fact, cataclastic processes take place to describe the true ductility in the mantle the law of Mises is used. The fluid saturation in the form of appropriate additives to the strength parameters of the medium is taken into account (in approximate form). Simulation is performed in a two-dimensional setting; the type of stress state is a flat deformation. The plates are characterized by lateral deviatoric stretching. The described structure of the stressed state is the a priori standard with which the results of modeling will be mainly compared.

    The initial data were used in [6–9], on the basis of which the geometry of the model was constructed and a generalized geophysical profile was created across the north of Honshu (Fig. 1). The model, at the initial moment of time, is a set of bodies consisting of square cells measuring 500×500 m, the total size of the model is 100 km vertically and 700 km lateral. The model is divided into 4 main parts (denoted by the color in Fig. 1), conditionally corresponding to the Upper, Middle and Lower crust and mantle part of the model. Separately, the water layer is introduced, as the real part of the model. Data on the state of stress are presented in [10]. We can single out the following most important regularities (note that it is specifically referred to deviator voltages). For the crust of the Japanese microcontinent, the geodynamic type of the stressed state is "lateral compression", as well as for the submerged cave west of the center of the depression. Moreover, if the axis of maximum compression $\sigma 3$ in the Pacific plate board is directed along the immersion of the crochet, then for the subcontinental part the axis $\sigma 3$ is inclined by $20-30°$ towards the trench. At the same time, for the part of the slab situated east of the trench, the lateral deviatoric stretching is characteristic. The described structure of the stressed state is the a priori standard with which the results of modeling will be mainly compared.

    As the main geodynamic sources of impact, the most classical ones are considered in our work: the effect of small-scale gravitational convection and the result of pressure from the side of the Pacific plate (the consequence of the play-tectonic processes formed under the action of general-body convection). In addition, an exogenous (as opposed to the previous endogenous) erosion source of exposure will be considered. Let us consider in more detail the modeling scheme for small-scale convection. The impact is modeled as the assignment of the corresponding vertical displacements at the base of the model: positive over the ascending and negative over the descending asthenospheric current. The descending current is given (for reasons of isostasy) under the most powerful bark (Honshu center), the ascending current is in the area of the Japanese depression. The dimensions and amplitudes of the effects of asthenospheric currents were selected in such a way that the result best suited the tectonophysical data on the stress state of the region described above. The simulation results are shown in Fig. 3.

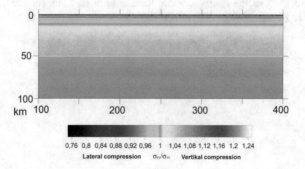

Fig. 3. Simulation results for a one-dimensional model for the formation of additional normal lateral compression stresses obtained as a result of removing the "erosion stress" of 1 km

The simulation results (Fig. 3) showed that the generalized model of the influence of mantle small-scale convection better corresponds to natural seismological data. The pressure from the spreading axis, anyway, cannot form the observed stressed state of the northern part of the Japanese subduction zone and should be considered in conjunction with other geodynamic processes to explain the formation of the current state of the region. The necessity of detailed account of erosion-accumulation processes, cardinally affecting the stressed state in the Upper and Middle crust, is also shown.

References

1. Ребецкий, Ю.Л., Полец, А.Ю.: Напряженное состояние литосферы Японии перед катастрофическим землетрясением Тохоку 11.03.2011. Геодинамика и тектонофизика, Т. 5, вып. 2, С. 469–506 (2014)
2. Ребецкий, Ю.Л., Сим, Л.А., Козырев, А.А.: О возможном механизме генерации избыточного горизонтального сжатия рудных узлов Кольского полуострова (Хибины, Ловозеро, Ковдор). Геология рудных месторождений, Т. 59, № 4, С. 263–280 (2017)
3. Ребецкий, Ю.Л.: О возможном механизме генерации в земной коре горизонтальных сжимающих напряжений. ДАН, Т. 423, № 4, С. 538–542 (2008)
4. Wilkins, M.L.: Computer Simulation of Fracture. Lawrence Livermore Laboratory, Rept. UCRL-75246 (1972)
5. Стефанов, Ю.П.: Некоторые особенности численного моделирования поведения упруго-хрупкопластичных материалов. Физ. мезомех, Т. 8, № 3, С. 129–142 (2005)
6. Nakamura, Y., Kodaira, S., Cook, B.J., Jeppson, T., Kasaya, T., Yamamoto, Y., Hashimoto, Y., Yamaguchi, M., Obana, K., Fujie, G.: Seismic imaging and velocity structure around the JFAST drill site in the Japan Trench: low Vp, high Vp/Vs in the transparent frontal prism. Earth Planets Space 66, 121–132 (2014)
7. Nishizawa, A., Kaneda, K., Oikawa, M.: Seismic structure of the northern end of the Ryukyu Trench subduction zone, southeast of Kyushu. Earth Planets Space 61, 37–40 (2009)
8. Nugraha, A.D., Mori, J., Ohmi, S.: Thermal structure of the subduction zone in western Japan derived from seismic attenuation data. Geophys. Res. Let. 37, L06310 (2010). https://doi.org/10.1029/2009gl041522
9. Iwasaki, T., Levin, V., Nikulin, A., Iidaka, T.: Constraints on the Moho in Japan and Kamchatka. Tectonophysics 609, 184–201 (2013)
10. Rebetsky, Y.L., Polets, A.Y., Zlobin, T.K.: The state of stress in the Earth's crust along the northwestern flank of the Pacific seismic focal zone before the Tohoku earthquake of 11 March 2011. Tectonophysics 685, 60–76 (2016)

Application of the Two-Frequency Radioholographic Method for Determinating the Location of Geoelectric Inhomogeneities in the Earth's Crust

V. A. Lubchich$^{(\boxtimes)}$ and V. F. Grigor'ev

Polar Geophysical Institute, Murmansk, Russia
lubchich@yandex.ru, valgri@pgi.ru

Abstract. The radioholographic method is the promising tool for solving problems of ore geophysics. Holographic reconstruction of geoelectric inhomogeneities in the earth's crust allows to effectively localize in space anomalous regions with high electrical conductivity, associated with local ore bodies, by using areal surface observations of the electromagnetic field. However, the analysis of previous experimental and model results demonstrated the possibility of "false" anomalies in the holographic reconstruction of the distribution of inhomogeneities. It was suggested to conduct observations at several frequencies as one of the ways of rejection of such "false" anomalies. So in 2017 extended field works were conducted at the site Loipishnjun in the Monchegorsk ore region by using the two-frequency radioholographic method. Results of field experimental studies have shown that the two-frequency modification of radioholographic method greatly improves the reliability of the holographic reconstruction of distribution of inhomogeneities in the earth's crust. The comparison of results of the holographic reconstruction at different frequencies allows one to reject "false" anomalies, and to identify anomalous zones, which can be associated with ore bodies.

The task of visualization of geoelectric inhomogeneities in the earth's crust often occurs when solving various geophysical problems, for example, when searching for ore bodies in the ore geophysics. In those cases, when it is enough to identify only the location of anomalously conducting zones in the earth's crust, the radioholographic method is the most rapid technique to determine the distribution of inhomogeneities. It is known that the superposition of two fields is recorded at observation points during electromagnetic sounding of the Earth. One is the primary field from the controlled source, which can be considered as reference wave. The other is the secondary fields caused by geoelectric inhomogeneities in the medium, which, in the holographic terms, can be referred to as the object waves. Thus, having results of measurements of the electromagnetic field over the studied area, one can reconstruct the distribution of the geoelectric inhomogeneities in the earth's crust.

The use of radioholographic method for the reconstruction of the location of sources of anomalous fields in the earth's crust is a new approach in the development of electromagnetic methods for exploration of mineral deposits. In the monograph [2] it

© Springer Nature Switzerland AG 2019
A. A. Zhamaletdinov and Y. L. Rebetsky (Eds.): SPS 2018, SPEES, pp. 27–33, 2019.
https://doi.org/10.1007/978-3-030-35906-5_5

was noted that the radioholographic reconstruction can be considered as a special case of the inverse scattering problem, because this method reconstructs locations of anomalous objects from measurements of electromagnetic fields that are scattered by these geoelectric inhomogeneities. The electromagnetic field at any point in space is the sum of the normal field determined by the controlled source in the absence of geo-electric inhomogeneities, and the anomalous field caused by currents induced in con-ducting regions of the earth's crust:

$$E(r) = E_0(r) + \int_V dr' G(r, r') j(r'),$$ (1)

where E - is the intensity of the total electric field, E_0 – is the intensity of the normal electric field, G – is the Green's function, j – is the current density producing the anomalous field. Integration goes over the domain containing anomalous sources.

On the other hand, in the solution of the forward problem the electromagnetic field at arbitrary point of space outside a closed volume may be determined from values of the electromagnetic field measured on the surface of this domain by using the Kirchhoff surface integral:

$$E(r) = \oint_S dS' [E(r')\nabla G(r, r') - G(r, r')\nabla E(r')]$$

For solving the inverse problem, it is required to determine the field inside this volume, i.e. in the domain accommodating sources of anomalous fields. This can be done by the Kirchhoff inversion formula:

$$E_H(r) = \oint_S dS' [E(r')\nabla G^*(r, r') - G^*(r, r')\nabla E(r')],$$ (2)

where G^* - is the complex conjugate Green's function. Formula (2) is the mathematical definition of the holographically restored field E_H. If the surface S is infinitely distant hemisphere and G – is the Green's function for the free space, the formula for the holographic reconstruction of the field is [2]:

$$E_H(\rho, z) = \frac{1}{2\pi} \int_\Sigma d\rho' E(\rho', z') \frac{\partial}{\partial z} \frac{exp(-ik\sqrt{|\rho - \rho'|^2 + (z - z')^2}}{\sqrt{|\rho - \rho'|^2 + (z - z')^2}},$$ (3)

where $k = \sqrt{i\omega\mu\sigma}$ – is the wave number for the lower half-space, ω – is the angular frequency of the field, μ – is the magnetic permeability, σ – is the electric conductivity of the earth's crust, ρ – is the horizontal projection of radius vector, i - is the imaginary unit. Integration goes over the earth's surface Σ, and areas where anomalous fields are sufficiently small are disregarded in the calculation.

By applying the Gauss theorem to the surface integral (2), we obtain the following relationship:

$$E_H(r) = \int_V dr[E\nabla^2 G^* - G^*\nabla^2 E].$$ (4)

The intensity of the electric field E and the complex conjugate Green's function G^* obey the Helmholtz equations:

$$\nabla^2 E = -k^2 E + j$$ (5)

$$\nabla^2 G^* = -k^2 G^* + \delta(r - r'),$$ (6)

where δ – the Dirac delta function. The substitution of formulas (5) and (6) into the Eq. (4) gives:

$$E_H(r) = E(r) - \int_V dr' G^*(r,r')j(r').$$ (7)

If we express the intensity of the total field E from formula (7) and substitute it into the Eq. (1), we find the relationship that links the holographic reconstruction of the field E_H with the unknown distribution of current density j, which generating anomalous fields:

$$E_H(r) = E_0(r) + 2i \int_V dr' j(r') Im\, G(r,r'),$$ (8)

where $Im\, G$ - is the imaginary part of the Green's function.

The grid approximation of the Eq. (8), when the lower half-space is divided into the cells with finite volumes, yields the system of linear equations with respect to unknown sources j. The determination of these values is the solution of the inverse problem of scattering.

In the derivation of formula (8), we considered electric components of the electromagnetic field; however, the similar relationship is also valid for magnetic components:

$$H_H(r) = H_0(r) + 2i \int_V dr' j_m(r') Im\, G_m(r,r').$$ (9)

But in this case the role of sources of the anomalous field is played by densities of fictitious magnetic currents j_m, which could be formally determined for the distribution of the real vortex currents density j in the following way:

$$j_m(r) = \nabla \times \int\limits_V dr'j(r')G(r,r').$$

Possibilities of the radioholographic method were experimentally tested on the Loipishnyun area of the Monchegorsk ore region. The Loipishnyun area is located on the southeastern slope of the Monchetundra massif, which is confined to the central part of the Pechenga-Varzuga rift structure. This rift zone cuts the Kola Peninsula from the northwest to the southeast and is represented by numerous intrusive massifs of basic and ultrabasic rocks, which accommodate the ore occurrences and deposits of copper-nickel, chrome, and titanium-magnetite ores and platinoid metals.

Geologically, the Monchetundra massif is a primarily stratified from dunites to leycogabbro lopolith [5]. The distinctive feature of this massif is the fact that, besides the main phase of the intrusion, also the additional norite-gabbronorite phase is identified within the eastern side of the massif. This area has an average thickness of up to 200 m. The additional phase includes a complex zone of development of sulfide-bearing rocks, which has the form of wedge. The base of the wedge-like block on the earth's surface is characterized by subvertical dip angles. This zone gradually narrows with the increasing depth and changes its orientation so that at the depths of about 1000 m its position is close to horizontal. Thus, the structure of the Monchetundra massif has a significant asymmetry, namely, the association of sulfide-bearing rocks mainly confined to its eastern side.

Intrusive massifs of the Pechenga-Varzuga rift structure are characterized by two types of the sulfide copper-nickel ore mineralization - magmatic (syngenetic) and metamorphic (epigenetic) ones. The drilling in the Loipishnyun area exposed the both types of mineralization. The magmatic type of mineralization is widely represented in rocks of the additional phase of intrusion: norites, gabbronorites, metagabbro. These rocks typically contain poorly interspersed sulfide mineralization in the amount of up to 1–3%. The boreholes also penetrated the bodies of massive sulfide ores related to the metamorphic type of mineralization. Since the metamorphic ore mineralization develops by redeposition of the primary magmatic ore material, it manly occurs in the zones of tectonic dislocations. Ore bodies of this type have higher concentrations of impregnated sulfides (up to 50–60%), usually combined with the nest-type mineralization and development of ore veins. The thickness of ore zones ranges from a few meters to the first tens of meters [4].

When analyzing results of previous works in 2012 [1], the possibility of "false" anomalies in holographic reconstruction of the inhomogeneity distribution was noted. As one of ways to reject such "false" anomalies it was proposed to undertake measurements at several frequencies. Therefore, in 2017, extended investigations were carried out on the Loipishnyun area of the Monchegorsk ore region by using the two-frequency radioholographic method for localization of ore zones in the space.

The measurements were carried out on ten profiles with a length of 900 m, the measurement step and the distance between profiles were 100 m. Thus, the uniform rectangular grid of 100 observation points was obtained. The spatial orientation of the grid was selected in such a way that one side of the survey area was perpendicular to the strike azimuth of rocks. In correspondence to this, the profiles were set up from the southeast to the northwest in the azimuth 312°.

The square ungrounded loop with a side length of 150 m was used as controlled source of the electromagnetic field. The center of the loop was located 400 meters South-East apart from edge of the survey area on profile 7. Sides of the loop were oriented along and across the direction of profiles. Harmonic signals at frequencies of 34 and 136 Hz were generated in the loop. The loop current was measured by the ammeter based on the Hall sensor and was 4–8 A on average. For obtaining the phase characteristics of the primary current, the signal was measured on a small segment of the loop (with a length of 1 m) and recorded to the data acquisition and storage system with precise GPS/GLONASS timing [3].

At points of observations, magnetic components of the field were measured by the three-component induction magnetometer with the digital data acquisition system similar to that used in the controlled-source equipment. Magnetic sensors had the mutually orthogonal orientation with the Y axis directed along profiles. Due to the precise timing of recorded signals, the transmitting and receiving equipment allowed to determine the absolute phase difference between magnetic field components and the current in the transmitting loop. Thus, in addition to amplitude values, we also obtained the distribution of phase characteristics of magnetic components over the survey area. It is the necessary condition for the holographic reconstruction of the distribution of geoelectric inhomogeneities in the earth's crust.

Based on measured values of amplitude-phase characteristics of magnetic field components on the surface of the survey area, values of holographic reconstructed magnetic field H_H in the lower half-space to a depth of $h = 1000$ m with a step of 100 m were calculated by the formula similar to the expression (3). The resistivity of the lower half-space was assumed to be 1000 $\Omega{*}$m, in accordance with the result of the previous electric resistivity prospecting in the Loipishnyun area by the method of mean-gradient electric profiling with measurement of induced polarization [5]. At the same points, we also calculated the normal field H_0 for the square ungrounded loop, located on the surface of a homogeneous half-space, and components of the magnetic Green tensor. After the grid approximation of the integral Eq. (9), the linear size of cubic cells was 100 m, the system of linear equations with respect to unknown densities of the magnetic current j_m at grid nodes was obtained. This system was solved by the standard Gauss technique.

In the case of a horizontal ungrounded loop, the excitation of the electric current in the earth's crust has the induction nature, i.e. eddy currents are predominantly sub-horizontal. Due to this, fictitious magnetic currents are mainly determined by their vertical component j_{mz}. Figure 1 presents maps of lines of equal amplitude for the vertical component of the magnetic current density j_{mz} for horizontal sections of the earth's crust $Z = 200, 300, 400$ and 500 m at frequencies 34 and 136 Hz, respectively.

The figure shows that at frequency of 34 Hz there are two intense anomaly within the range of coordinates $X = 100$–300 m, $Y = 400$–600 m, $Z = 300$–400 m and $X = 600$–800 m, $Y = 200$–500 m, $Z = 300$–500 m. At frequency of 136 Hz, the first anomalous region is observed almost in the same range of coordinates, however, its form is less vague. The second anomalous region at the cross section of the earth's crust $Z = 300$ m is not observed, the relatively weak anomalous zone can be seen on deeper horizons $Z = 400$–500 m within the coordinates $X = 600$–800 m, $Y = 200$–300 m.

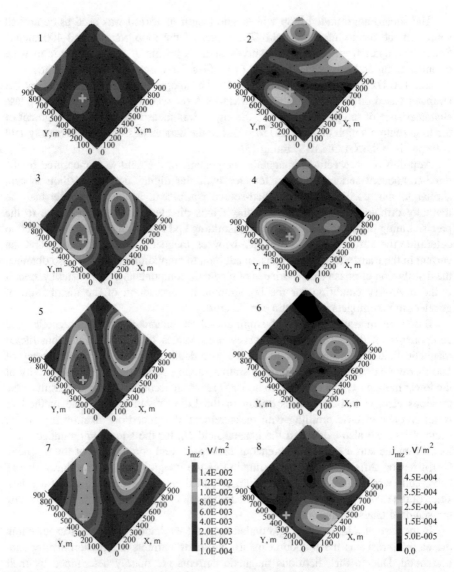

Fig. 1. Contour map of amplitude of the vertical component of magnetic current density j_{mz}: 1 - horizontal cross section of the earth's crust $Z = 200$ m at frequency of 34 Hz; 2 - at frequency of 136 Hz; 3 - section $Z = 300$ m at frequency of 34 Hz; 4 - at frequency of 136 Hz; 5 - section $Z = 400$ m at frequency of 34 Hz; 6 - at frequency of 136 Hz; 7 for cross-section $Z = 500$ m at frequency of 34 Hz; 8 - at frequency of 136 Hz. The green cross indicates the point of intersection of the borehole C-1720 at this horizon

The first anomalous region with high electrical conductivity has geological explanation. In this area there is the deep well C-1720, the length of the wellbore is 502.7 m [4]. The borehole is located within the secondary sulfide-bearing phase of the Monchetundra massif and is drilled across the strike of rocks. This borehole exposed both rocks with poor impregnations of sulfide ore mineralization and ore zones of metamorphic type with rich sulfide mineralization. The interbeds with the redeposited copper-nickel sulfide mineralization were detected in intervals 218–219, 265–270 and 360–365 m and were confined, as a rule, to zones of tectonic faults. In the figure, the location of the borehole for various horizontal sections of the earth's crust is indicated by the green cross. By comparing these data with the holographic reconstruction of geoelectric inhomogeneities in the earth's crust, we can see that this highly conductive anomalous area corresponds to the position of metamorphic ore zones with rich sulfide mineralization.

The second anomalous area is not certified by drilling wells. The comparison of results of the holographic reconstruction of the distribution of geoelectric inhomo-geneities for different frequencies allows us to conclude that some zone with high electrical conductivity can be located on deeper horizons of the earth's crust. For the horizontal section $Z = 300$ m at frequency of 34 Hz, this anomaly can be considered as "false", since it does not detected at frequency of 136 Hz.

Thus, results of the field experimental work in the Loipishnyun area allow us to conclude, that the radioholographic method is the promising tool for solving problems of the ore geophysics. The holographic reconstruction of geoelectric inhomogeneities in the earth's crust allows to effectively localize in space anomalous regions with high electrical conductivity, associated with local ore bodies, by using areal surface observations of the electromagnetic field. However, for a more reliable interpretation of results of the holographic reconstruction of the distribution of inhomogeneities in the earth's crust, studies should be carried out in a two-frequency version of the method. The comparison of results of the holographic reconstruction at different frequencies allows one to reject "false" anomalies, and to identify anomalous zones, which can be associated with ore bodies.

Acknowledgments. The study is executed at financial support of RFBR and the government of the Murmansk region (project No. 17-45-510956).

References

1. Lyubchich, V.A.: Application of the radio holographic method in the prospecting for local ore bodies. Izv. Phys. Solid Earth **51**(2), 290–299 (2015)
2. Tereschenko, E.D.: The radioholographic method for the study of ionospheric inhomo-geneities. Apatity, 99 p. (1987). (in Russian)
3. Filatov, M.V., Pilgaev, S.V., Fedorenko, Yu.V.: A four-channel 24-bit analog-to-digital converter synchronized with the universal-time clock. Instrum. Exp. Tech. **54**(3), 361–363 (2011)
4. Sholohnev, V.V., Polyakov, I.V., et al.: Report on results of the exploration of sulfide copper-nickel ores and other minerals in the contact zone of the Monchegorsk pluton with the Monchetundra massif in 1994–98 (the Loypishnjun site), Monchegorsk (1998). (in Russian)
5. Sholohnev, V.V., Pustovoytov, V.S., et al.: Report on results of the exploration of chrome ores and other minerals in the southern part of the Monchegorsk ore region in 2001–2003, Monchegorsk (2003). (in Russian)

Structures of the Earth Crust and Their Interaction with Srresses. Geomechanics and Tectonophysics Data

Yu. L. Rebetsky[(✉)]

Schmidt Institute of Physics of the Earth, Moscow, Russia
reb@ifz.ru

Abstract. Russian geomechanics made an important contribution to the development of ideas about the deformation properties of rocks in the litho-sphere (Nikolaevsky 1979, Nikolaevsky 1996, Nikolaevsky 2006, Nikolaevsky and Sharov 1985) in the second half of the 70's - the beginning of the 80's of the last century.

Russian geomechanics made an important contribution to the development of ideas about the deformation properties of rocks in the lithosphere (Nikolaevsky 1979, Nikolaevsky 1996, Nikolaevsky 2006, Nikolaevsky and Sharov 1985) in the second half of the 70's - the beginning of the 80's of the last century. In my report, the development of the ideas of VN Nikolaevsky will be made on the basis of the new results of geomechanical experiments and tectonophysical data on the current tense state of various seismically active regions.

VN Nikolaevsky suggested that the bottom of the crust is the boundary of the change in the elastic-brittle behavior of rocks to an elasto-plastic one, in which irreversible deformations arise due to displacement of dislocations, crushing and sliding along grain boundaries. This was the result of the interpretation of laboratory experiments at temperatures of 500–600° and pressures of 10–15 kbar, showing the closure under these conditions in the rock cracks, In this case the boundary of M should be considered as a barrier for the penetration of meteoric water into the mantle (see also (Ivanov 1990)).

In dry conditions, strength is the ultimate level of deviator stresses in rocks which one would be determined by the level of the lithostatic pressure, which under these conditions creates increased values of friction forces on the surface of cracks. In real cortical conditions the strength of the fractured massifs, though exceeds the strength of the internal cohesion (sometimes several times), but the fluid pressure in the fractured pore space reduces the role of the lithostatic rock pressure. In the mantle lithosphere, water is found in the mineral composition of the rocks mainly in the form of a hydroxyl group, and here the level of deviator stresses is directly determined by the yield strength. Thus, the boundary the M is the boundary of the change in the mechanical behavior of rocks and the boundary of the change in the mineral composition of the lithosphere.

Probably, Nikolaevsky (1980) first used the Mohr diagram (Fig. 1a) in the analysis of the process of brittle fracture in the earth's crust with the release of a brittle fracture

A. A. Zhamaletdinov and Y. L. Rebetsky (Eds.): SPS 2018, SPEES, pp. 34–42, 2019.
https://doi.org/10.1007/978-3-030-35906-5_6

zone - the rock dilatancy zone. Later in work (2006) he introduced on this diagram both areas of fragmentation and mylonitization - a zone of compaction (see, for example, (Stefanov 2008)). He also first tried to explain from the position of geomechanics the waveguides isolated in the middle of the seventies in the middle crust, as well as listric faults (Suess 1909) that flatten out at different levels (boundary M, wave channel) (Ogarinov 1974, Pavlenkova 2000).

It is believed that the lower edge of a listric faults is a zone of global horizontal detachment (see, for example, (Yudakhin 2003)). The actual data on the properties and state of the rock derived from the depths M on the surface, show that they have always been represented by tectonic breccias and milonites separating the fragile part of the earth's crust from the mantle deformed plastically.

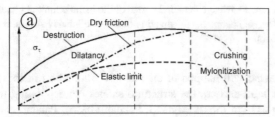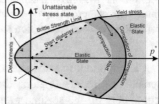

Fig. 1. Crustal stress states on the Mohr diagram: (a) on the work of Nikolaevsky (2006) (the figure is shown unchanged); (b) tectonophysical concepts of stressed states in the crust (Rebetsky 2008a, b): 1- detachment area; 2 – dilatancy area; 3 - compaction area

VN Nikolaevsky considers the middle crust as the place of the most pronounced cataclastic fracture (zone of bright dilatancy), which in the presence of fluids con-tributes to the formation of seismic waveguides (Elansky 1964). The scale of the fractures and cracks here corresponds to the megascopic level, and at a certain size of the averaging window the deformation behavior of the rocks of the middle crust can be considered as pseudo plastic. VN Nikolaevsky believes that in the lower crust, where the level of lithostatic pressure is much higher, rocks are defragmented to a milonitstate and fracture deformations here manifest themselves as pseudoplastic with a smaller averaging scale.

Viktor Nikolaevich suggested the hypothesis that in the upper crust fractures are dominated by subvertical cracks (in fact, cracks of separation caused by the splitting effect), which flatten out in the middle crust (Fig. 2a). However, he believed that the tectonic stresses of maximum compression near the top of the crust are subvertical. He explained the smoothing of faults with depth closer to the middle crust by increasing the lateral - horizontal reduction, which in the experiment corresponded to shear cracks composing angles of 25–40° with axial compression. VN Nikolaevsky attributed the change in the orientation of the maximum compressive stresses to the subhorizontal (Nikolaevsky and Sharov 1985), the next level of even more flattening of the faults, which occurs as we approach the top of the middle crust and the bottom crust. Similar views for the interpretation of seismic profiles were also used in (Puchkov and Svet-lakova 1993, Puchkov 2000).

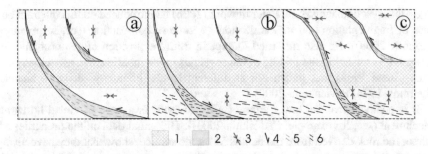

Fig. 2. The relationship between the stressed state and position in the crustal faults and reflecting areas: (a) by VN Nikolaevsky (Nikolayevsky and Sharov 1985); (b, c) according to the modern tectonophysical representations presented in this paper for the modes (a) of horizontal extension and (b) horizontal compression. 1 - wave channel, 2 - zone of the listratic fault, 3 - axes of maximum compression, 4 - directions of shears, 5 - localization bands of dilatancy shears, 6 - localization bands of compaction shears

It is important to understand what will happen in the zone of articulation of these opposite displacements along the fault and are the structural shapes corresponding to them visible on the seismic profiles? On the other hand, Patalah (1986), Patalah and Khrychev (1988) showed that listric faults, at least in the upper part of them, can be faults in the extension (normal faults) or compression (inversion faults). They believe that the platform is a extension structure, and in the folded areas - the compression structure. Thus, the relationship between the stress state and the geometry and structure of the crust is more complex. In addition, data on the seismicity of the crust show that, in the presence of wave channel, earthquakes occur in the upper crust, but ignore the wav channel. This may indicate that there is no necessary level of deviator voltages for large-scale destruction.

Hypothesis VN Nikolaevsky on the orientation of the axes of the principal stresses in the crust disagreed with the known in situ measurements of nature stresses (Markov 1980, Markov 1985). In these papers it was shown that the geodynamic type of the stressed state is related to the nature of the motion of the surface. In the regions of uplift (mountain uplifts) and prolonged denudation (shields) in a larger number of cases (>65% points of the observations), the horizontal compression mode takes place. Conversely, in the bark of active deflections (sedimentary basins) and plates, as a rule (>70% points of the observations) horizontal tension occurs. If in the zones of different intensity of deflections the regime of the stressed state with depth does not change, then the regions of uplifts with depth decrease gradually and already at depths of the order of 3–5 km are almost comparable with the magnitude of the vertical compressive stresses. In the works (Rebetsky 2008a, b) on again after (Voigth and Pierre 1974, Goodman 1987), attention was concentrated to the fact that horizontal compression in the upper crust may be associated with erosion-denudation processes that lead to the surface of rocks that have undergone a supercritical state in depth and in which there are additional stresses of horizontal compression (not with the push of the "Indian plate"). This additional compression is retained for the most part even when the rock leaves closer to the surface. With regard to stress problems in the Kola Peninsula, it was shown that the level of denudation of the surface of about 2, 5 and 11 km, which took place here for the times of 50-150-250 Ma, respectively, may well explain the can to static level of horizontal compression stresses of 50–150 bar (Rebetsky et al. 2017).

On the other hand, subvertical faults are most often observed in sedimentary basins (Kunin et al. 1988). In mountain areas, seismic data (Emanov et al. 2005, 2006, 2010) quite clearly show the submergence of faults under the ridges, which only gradually converts to a steeper depth (see Fig. 2c) (Abdrakhmatov et al. 2001).

Wet possible to give a different interpretation on the basis of experimental data obtained in the last 20 years on the destruction of rocks at high pressures make. In particular, from the geomechanical experiment (Chemenda et al. 2007, Chemenda et al. 2011) it follows that with increasing compression of the samples, the angle of internal friction determining the position of the cleavage site with respect to the axial pressure gradually decreases. This determines the change in the orientation of the cleaved cracks with respect to the axial pressure (Fig. 3).

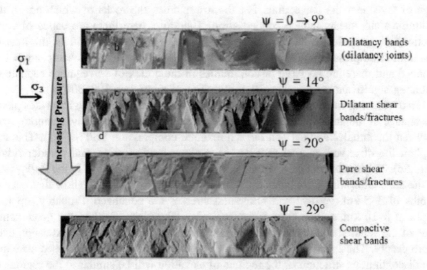

Fig. 3. Results of experiments for brittle fracture of artificial stone in the face of increasing compression (Figure from the work (Chemenda et al. 2011))

Geomechanical experiments, for example (Chemenda et al. 2007, Chemenda et al. 2011), also showed the presence of not only dilatancy areas in the Mohr diagram, where there is an irreversible shear and increase in volume, but also compaction (Fig. 1). Compaction in experiments occurs at very high axial pressure, performed under conditions of rigid lateral fixation of the samples. In this case, the crushing of the projections of the grains begins and the compaction of the rock occurs due to their rotations and displacements. In this case, bands of compaction appear in the sample, similar to the dilatancy bands, with the difference that they are compaction rather than dilatancy. These bands can occur under different conditions, i.e. with a different combination of the level of all-round compression and shear stresses. In this case, their inclination with respect to the axial pressure in the sample is different. With a high level of stress (deviator stresses), the slope of the compaction bands can reach 50–60°, and at a low level (80–90°) (suborthogonal to the axial pressure).

It follows that in the upper part of the section for the horizontal extension mode (Fig. 2b), the faults can indeed have a fairly steep dip, sometimes close to the sub-vertical (angles of immersion of about 80–90°), as noted in (Nikolayevsky and Sharov 1985). With depth from the action of mass forces, not only the level of vertical compression of the P_V increases, but also the level of horizontal compression of rocks the P_H. At the same time, their ratio P_H/P_V increases ($\rightarrow 1$). In such conditions, according to experiments, there should be a gradual decrease in the angle of inclination of cracks and faults. At the same time, fragile shifts accompanied by a dilatancy mechanism (decompaction) can not deviate from the axis of maximum compression to angles greater than 35–45°. Such angles of inclination are sufficient to form a listratic type of fault near the waveguide. For the lower crust, the rocks of which are in the milonitic state, shears induce condensation. Therefore, here there are bands of compaction, whose poles at angles of 10–30° deviate from the direction of the axis of maximum compression. With PV/PH close to unity, the level of deviator stresses is reduced and there are no large discontinuities in the cortex of waveguides capable of initiating significant earthquakes (discontinuities of more than 100–300 m).

In the crust of mountain rises where the horizontal compression regime takes place, the change in the angles of fault submersion can be associated with the changing ratio between the smaller vertical and large horizontal compression P_V/P_H ($\rightarrow 1$). This can explain the observed gradual increase in angles of submerging faults under ridges (Fig. 2c). The in-situ data (Fig. 4) and our calculations of the change in the P_V/P_H ratio in the areas of uplifts accompanied by erosion (Rebetskii 2008a, b) show that even at depths of 3–5 km, vertical and horizontal stresses are equalized. Probably, up to a depth of 8–10 km, a state close to isotropic can take place. Further, the geodynamic type of the stressed state changes from the horizontal to the horizontal extension even more deeply. Thus, starting from the depths of the middle crust, the stressed state and the discontinuous structure of the regions of elevation will be similar to the regions of horizontal extension (Fig. 2b).

Proposed by VN Nikolaevsky interpretation of the mechanical state of the upper part of the lithosphere and the following estimates of the strength (Fig. 5a) of the middle crust differ somewhat from the corresponding models developed in the works of western researchers (Fig. 5b) (Ranalli and Murphy 1987). Here the middle part of the crust was characterized from the position of a truly plastic deformation - intra-crystalline dislocation creep. The boundary of M was also considered to be a transition from a fractured flow to a true plastic one.

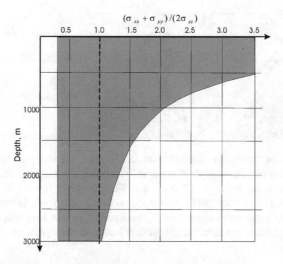

Fig. 4. The ratio of vertical and horizontal compressive stresses in the depths of uplifts. Figure from the work (Brady and Bzown 2004) with changes

It is important to note that initially in (Sibson 1974, Ranalli and Murphy 1987, Ranalli 1995, etc.) it was believed that the fluid pressure was distributed in depth according to the hydrostatic law (the weight of the liquid column). Proceeding from this, estimates of the level of shear stresses in the depth of the crust were made. Thus, in the areas of the crust of horizontal compression at a depth of 30 km, the maximum shear stress was found to be about 5 kbar, while the average for the stress crust was estimated at 2.5 kbar. In the works of VN Nikolaevsky did not specify the magnitude of the fluid pressure.

In the works of Ivanova (1992, 1994) and IG Kissin (1996) specially paid attention to the possibility of achieving fluid pressures above the hydrostatic values. In particular, from the sharp change in the permeability of rocks at the boundary between the upper and lower crust, there is a over hydrostatics fluid pressure in the roof of the waveguide. Since they should increase with depth according to the law of hydrostatics (the waveguide is a closed system and the increase in depth occurs due to the weight of the fluid in the fractured pore system), then the fluid pressure at the bottom of the waveguide is lower than the lithostatic (see Fig. 5c).

Developing these ideas, we believe that in the lower crust near the soles, due to the closure of pores and cracks, the fluid pressure should approach lithostatics. Therefore, the frictional forces on the cracks approach zero, and the strength of the rocks in the lower crust tend to their cohesion strength. Thus, the lower crust, with the exception of a small section near its roof, has strength even lower than in the seismic channel. The difference between the state of the lower crust and the middle (seismic channel) consists in a denser packing of the rock and a smaller volume of the fractured-pore space, which is manifested in an increase in the velocity of seismic waves with depth.

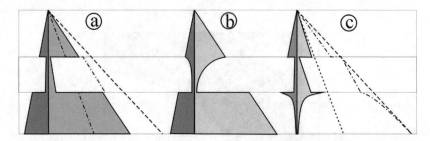

Fig. 5. The level of maximum shear stresses in the crust according to the model: (a) VN Nikolaevsky (in the middle crust fluid pressure was taken according to the model of Kissin (2015)); (b) scheme according to (Ranalli and Murphy 1987 and etc.); (c) according to tectonophysical data (Rebetsky 2015), developed in this work. Short dashed line is the fluid pressure distributed in the depth of the crust by the hydrostatic law; long dotted line - lithostatic pressure; bar-dotted line - the assumed law of distribution of fluid pressure

From the data of geophysical profiles it is known that in the lower part of the cortex there are a large number of reflecting sites (Fig. 2c), which are associated with the phenomenon of global shear along the bottom of the crust (Yudakhin 2003). However, the relationship between the compaction bands and axial pressure directions obtained in the geomechanical experiment makes it possible to give a different interpretation to these areas. These sites should be considered as compaction bands - compression with a slight shear arising in the lower crust under conditions of a high level of vertical compressive stresses and with a high level of lateral restraint.

The equations connecting angles of fault submergence for two different types of geodynamic regimes (tension and compression) are proposed with stress values. "Strength envelopes" for the modes of horizontal compression and tension are calculated.

References

Abdrakhmatov, K.E., Weldon, R., Thompson, S., Burbank, D., Rubin, C., Miller, M., Molnar, P.: Origin, direction and velocities of contemporary compression of the central Tien Shan (Kirghizia). Geol. Geophys. **342**(10), 1585–1609 (2001). (in Russian)

Adushkin, V.V., Rodionov, V.N.: On the nature of mechanical motion in the Earth's subsoil. Phys. Earth (2), 88–91 (2005). (in Russian)

Brady, B., Bzown, E.: Rock Mechanics for Underground Mining, 3rd edn., 688 p. Kluwer Academic Publishers (2004)

Chemenda, A.I., Balas, G., Soliva, R.: Impact of a multilayer structure on initiation and evolution of strain localization in porous rocks: field observations and numerical modeling. Tectonophysics **631**, 29–36 (2014)

Chemenda, A.I., Nguyen, S.-H., Petit, J.P., Ambre, J.: Mode I cracking versus dilatancy banding: experimental constraints on the mechanisms of extension fracturing. J. Geophys. Res. **116**, B04401 (2011)

Chemenda, A.I.: The formation of shear-band/fracture networks from a constitutive instability: theory and numerical experiment. J. Geoph. Res. **112**, B11404 (2007)

Elanskiy, L.N.: Deep Earth's hydrosphere. Tr. Kuibyshev NIIP, no. 26. pp. 118–152 (1964). (in Russian)

Emanov, A.F., Emanov, A.A., Filina, A.G., Leskova, E.V.: Spatio-temporal features of the seismicity of the Altai-Sayan folded zone. Phys. Mesomech. **8**(1), 49–64 (2005). (in Russian)

Emanov, A.F., Emanov, A.A., Filina, A.G., Leskova, E.V., Kolesnikov, Y.I., Rudakov, A.D.: General and individual in the development of aftershock processes of the largest earthquakes in the Altai-Sayan mountain region. Phys. Mesomech. **9**(1), 33–43 (2006). (in Russian)

Emanov, A.F., Emanov, A.A., Leskova, E.V.: Seismic activization in the Belino-Busiging zone. Phys. Mesomech. **13**(1), 72–77 (2010). (in Russian)

Goodman, R.: Mechanics of Rock, 232 p. Stroyizdat, Moscow (1987). (in Russian)

Ivanov, S.N.: On rheological models of the earth's crust; critical consideration. Publishing House of the Ural Branch of the Russian Academy of Sciences, Ekaterinburg (1998). (in Russian)

Ivanov, S.N.: Probable nature of the main seismic boundaries in the Earth's crust of the continents. Geotectonics (3), 3–11 (1994). (in Russian)

Ivanov, S.N.: Rheological zoning of the lithosphere, the nature and significance of the K1 boundary. Metamorphogenic Metallogeny of the Urals, 44 p. UB RAS, Ekaterinburg (1992). (in Russian)

Ivanov, S.N.: The separator (on the nature and significance of the geophysical boundary of K1). Dokl. AN SSSR. **311**(2), 428–431 (1990). (in Russian)

Kadik, A.A., Lukanin, O.A.: Degassing of the upper mantle during melting, 97 p. Science, Moscow (1986). (in Russian)

Kissin, I.G.: Fluids in the Earth's Crust. Geophysical and Tectonic Aspects, 2nd edn. 328 p. Science, Moscow (2015). (in Russian)

Kissin, I.G.: Metamorphogenic dehydration of crustal rocks as a factor of seismic activity. Doklady RAN **351**(5), 679–682 (1996). (in Russian)

Kunin, N.Ya., Ioganson, L.I., Afonsky, M.N., Abetov, A.E., Daukaev, S.Zh.: Continental basins of Central and East Asia. Regularities of deep structure and development, 168 p. Izd. IPE RAS, Moscow (1988). (in Russian)

Letnikov, F.A., Karpov, I.K., Kiselev, A.I., Shkandria, B.O.: Fluid regime of the earth's crust and upper mantle, 216 p. Science, Moscow (1977). (in Russian)

Levin, B.V., Rodkin, M.V., Sasorova, E.V.: On the possible nature of the seismic boundary at a depth of 70 km. Dokl. RAS. **414**(1), 101–104 (2007). (in Russian)

Markov, G.A.: On the distribution of horizontal tectonic stresses in zones of uplifts of the Earth's crust. Eng. Geol. (1), 20–30 (1980). (in Russian)

Markov, G.A.: The patterns of tectonic stress distribution in the upper part of the earth's crust. New data and practical applications. Interrelation of geological-tectonic structure, properties, structural features of rocks and manifestations of excessive tension. Apatity: Col. Phil. Of the USSR, pp. 72–84 (1985). (in Russian)

Model analysis of intraplate mantle-crust ore-forming systems, 409 p. Publishing house of the SB RAS, Novosibirsk (2009). (in Russian)

Nikolaevsky, V.N.: Cataclastic destruction of crustal rocks and anomalies of geophysical fields. Phys. Earth (4), 41–50 (1996). (in Russian)

Nikolaevsky, V.N.: Dilatansiya and the theory of the earthquake focus. In: Advances in Mechanics, vol. 3, no. 1, pp. 71–101 (1980). (in Russian)

Nikolaevsky, V.N.: Geomechanics and Fluid Dynamics, 446 p. Nedra, Moscow (1996). (in Russian)

Nikolaevsky, V.N.: Mechanics of geomaterials and earthquakes. Itogi Nauki i Tekhniki VINITI. Ser. Meh.def.tv.teela, vol. 15, pp. 817–821 (1983). (in Russian)

Nikolaevsky, V.N.: Overview: earth's crust, dilatancy and earthquakes. Mechanics of the earthquake focus, pp. 133–215. Mir, Moscow (1982). (in Russian)

Nikolaevsky, V.N.: The boundary of Mohorovicic as the ultimate depth of the delicate-dilatancy state of rocks. DAN SSSR **249**(4), 817–820 (1979). (in Russian)

Nikolaevsky, V.N.: Fracture of the Earth's crust as its genetic sign. Geol. Geophys. **47**(5), 646–656 (2006). (in Russian)

Nikolaevsky, V.N., Sharov, V.I.: Rifts and rheological stratification of the earth's crust, Izv. AN SSSR. Phys. Earth (1), 16–28 (1985). (in Russian)

Ogarinov IS Deep structure of the Urals, 79 p. Science, Moscow (1974). (in Russian)

Patalaha, E.I.: To the problem of listric faults. Izv. AN SSSR. Ser. geol. (11), 113–127 (1986). (in Russian)

Pataloha, E.I., Khrychev, B.A.: Listric faults in folded regions. Geotektonika (4), 8–19 (1988). (in Russian)

Pavlenkova, N.I.: The main results of deep seismic sounding for 50 years of research. Regional geology and metallogeny, no. 10, pp. 12–20 (2000). (in Russian)

Puchkov, V.N.: Paleogeodynamics of the Southern and Middle Urals, 146 p. Publishing House of Dauria, Ufa (2000). (in Russian)

Puchkov, V.N., Svetlakova, A.N.: The structure of the Southern Urals along the Trinity Profile of the DSS. Dokl. AN SSSR **333**(3), 348–351 (1993). (in Russian)

Ranalli, G.: Rheology of the Earth, 2nd edn., 413 p. Chapman and Hall, London (1995)

Ranalli, G., Murphy, D.C.: Rheologycal stratification of the lithosphere. Tectonophysics **132**, 281–295 (1987)

Rebetsky, Yu.L.: On the peculiarities of the stressed state of the crust of intracontinental orogenes. Geodyn. Tectonophys. **6**(4), 437–466 (2015). (in Russian)

Rebetsky, Yu.L.: The mechanism of generation of residual stresses and large horizontal compressive stresses in the earth's crust of intraplate orogens. Problems of tectonophysics. To the 40th anniversary of the creation of M.V. Gzov Laboratory of Tectonophysics at the Institute of Physical Chemistry of the Russian Academy of Sciences, pp. 431–466. Izd. IPE RAS, Moscow (2008a). (in Russian)

Rebetsky, Yu.L.: The mechanism of generation of tectonic stresses in the regions of large vertical movements. Phys. Mesomech. **11**(1), 66–73 (2008b). (in Russian)

Rebetsky, Yu.L., Sim, L.A., Kozyrev, A.A.: On the possible mechanism of generation of excessive horizontal compression of ore nodes of the Kola Peninsula (Khibiny, Lovozero, Kovdor). Geol. Rud. The deposit., vol. 59, no. 4, C. 263–280 (2017). (in Russian)

Sibson, R.H.: Frictional constraints on thrust, wrench and normal faults. Nature **249**(5457), 542–544 (1974)

Stephanov, Yu.P.: Numerical modeling of processes of deformation and destruction of geological media. Abstract of diss. to the soisk. uch. Art. Ph.D. (in Russian)

Suess, E.: The face of the Earth. Oxford, vol. 1Y (1909)

Voigth, B., St Pierre, B.H.P.: Stress history and rock stress. In: Proceedings of Third International Congress of the International Society for Rock Mechanics, Denver, vol. 2, pp. 580–582 (1974)

Yudakhin, F.N.: About some common features of the deep structure and geodynamics of platforms and mountain-folding regions. Geodynamics and geoecological problems of high mountain regions. In: Materials of the Second International Symposium, Bishkek, 29 October–3 November 2002, pp. 47–68. Izd. Penthouse, Bishkek-Moscow (2003)

On the Study of Lithosphere Temperature from Electromagnetic Sounding Results

A. N. Shevtsov[1(✉)] and A. A. Zhamaletdinov[1,2]

[1] Kola Science Center, Geological Institute, Apatity, Russia
anshev2009-01@rambler.ru, abd.zham@mail.ru
[2] Saint-Petersburg Branch of IZMIRAN, Saint Petersburg, Russia

Abstract. The article presents a method of calculating of the lithospheric temperature dependence vs depth on the base of a-priori information of electrical resistivity change vs depth (based on the results of electromagnetic soundings), the composition of rocks at depth (based on geological assumptions) and with the use of laboratory data of electrical properties of dry rocks at high temperature. It is assumed that at a depth of more than 10–15 km, the resistivity of the lithosphere is mainly determined by the composition of the rocks and the effect of temperature. The solution of the problem is illustrated by the example of calculations performed for two types of geoelectrical section of the Fennoscandian shield - "normal" and "anomalous". Information on geoelectrical sections was obtained from the results of electromagnetic soundings with industrial power lines (FENICS experiment). The deep geological structure of the lithosphere is assumed to be the same in "normal" and "anomalous" areas. The maximum temperature difference within these two areas was about 70 °C in the depth range of 30–60 km.

Keywords: Electrical sounding · Lithosphere · Conductivity · Temperature · Fennoscandian shield

1 Introduction

The task of thermal regime study at depth is one of the fundamental problems of the Earth physics. Temperature, acts as the most important physical factor determining the electrical conductivity of mining rocks at depth (Keller 1966; Keller 1981; Waff 1974; Zhamaletdinov 1982; Spies and Frischknecht 1992; Zhdanov and Keller 1994; Vanyan 1997; Spichak and Zakharova 2015), as well as viscoelastic and rheological characteristics of rocks. (Fernandez and Ranalli 1997; Goes et al. 2000; Cammarano et al. 2003; Glaznev 2003; Brantut et al. 2012). A detailed analysis of well-known indirect geothermometers and a method of extrapolation to depth of temperature measurements in wells based on a neural network algorithm using geo-electrical data is developed in (Spichak et al. 2015). On the basis of information about the geoelectric section, it is possible to calculate the temperature profile at depth using the exponential law of Ioffe (1974) and relying on the known dependences of the specific conductivity of rock samples on temperature based on published laboratory data (Parkhomenko 1965; Parkhomenko and Bondarenko 1972; Chelidze et al. 1979). In this paper, an algorithm

© Springer Nature Switzerland AG 2019
A. A. Zhamaletdinov and Y. L. Rebetsky (Eds.): SPS 2018, SPEES, pp. 43–53, 2019.
https://doi.org/10.1007/978-3-030-35906-5_7

is created in which thermal estimates are constructed using an approximating piecewise continuous function. In this case, the effect of pressure on the specific resistance of rocks is not taken into account. Let the geoelectric section be known with some accuracy according to the CSAMT results. In addition, temperature dependences of the specific electrical resistance of the main types of dry rocks—granites, diabases, basalts, peridotites, eclogites, etc.—are known from the laboratory data. Further, the developed algorithm can be used to the two types of problems solution.

1. Let it be known (according to petrology data) in what sequence and in what ratios different types of rocks follow each other in depth and also it is known the change in electrical conductivity at a depth from the results of electromagnetic sounding. In this case, using laboratory data on rock resistivity on samples, the task can be set to determine the temperature distribution at depth for a given geoelectric section and for a given type of geological structure of the earths crust
2. Let it be known the temperature distribution at depth (from the data of measurements of heat flux and seismic prospecting) and also it is known the change in electrical conductivity at depth (from results of electromagnetic sounding). In this case, the task can be set to determine the parameters of the geological structure of the earth's crust (the approximate composition of rocks at depth) for a given geoelectrical section and for a given temperature distribution.

In the present work, the problem of the first type is considered on the example of two types of geoelectric sections («normal» and «anomalous») obtained under the conditions of the same geological structure of the earth's crust (as we suppose). Geoelectric sections were obtained from the results of soundings on the Fennoscandian shield (experiment FENICS) (Zhamaletdinov et al. 2015).

2 Physical Foundation of Geothermal Calculations in Geoelecrics

The relationship of the electrical resistivity of rocks ρ_r with the electrolyte (fluid) resistivity (ρ_f) and with the porosity of rocks in relative units (p) is determined by Eq. (1) following from Archie's law (Archie 1942)

$$\rho_r = \rho_f/p^n, \tag{1}$$

Where n is a dimensionless coefficient varying from 1.5 to 2 depending on connectedness degree of electrical channels. For example, if we have a porosity of 5% ($p = 0.05$) and a fluid resistivity ρ_f of 250 Ω m, the resistivity of rock ρ_r, depending on the degree of connectivity of the channels, can vary around 10^4–10^5 Ω m. With depth, the porosity of rocks decreases under the action of lithostatic pressure (Zharikov et al. 2003), which leads to a decrease in humidity and an increase in the resistivity of rocks. But at high temperature (about 500 °C) and high pressure – in supercritical state a conductivity of fluids increases to maximum value that is about 1.5 order compare to conductivity at 18 °C (Quist and Marshall 1968, Quist and Marshall 1969; Vanyan 1984; Dai et al. 2015; Kummerow and Raab 2015). At depths of tens of kilometers and more, the resistivity of rocks is determined not so much by the presence of moisture as by the properties of the mineral skeleton and the effect on its electrical conductivity of temperature, increasing with depth with an average geothermal gradient around 10–30 °C/km. The dependence of electrical resistivity of silicate rocks on temperature in this case can be estimated by exponential law Ioffe (1974) represented by Eq. (2)

$$\rho = \rho_0 \cdot e^{E_0/kT}, \tag{2}$$

where ρ_0 is the pre exponential coefficient that is numerically equal to the specific resistivity of rock at temperature $T \rightarrow \infty$; E_0 is the activation energy of current carrying ions (in electron-volts, 1 eV = $1.602 \cdot 10^{-19}$ J); k_B is the Boltzmann constant ($k_B = 1.38 \ 10^{-23}$ J/K); T is the absolute temperature (K) on the Kelvin scale. If to take logarithms from the left and right parts of the above Eq. (2), we will find:

$$\log \rho = \log \rho_0 + 0.434 \cdot \frac{E_0}{k_B T} \tag{3}$$

Since the pre exponential coefficient is a constant determining the type of electricity carriers, it is customary to depict the resistivity dependence on temperature on a linear-logarithmic scale, as it is presentsed in Fig. 1. Which shows the average diagrams of the dependence of the specific resistance of acidic and basic-ultrabasic rocks of the Fennoscandian shield vs temperature T (Parkhomenko 1965). One can see that the resistivity of rocks decreases with increasing temperature, on average, by one order per every 100°. Against the background of an approximately linear dependence of lg ρ vs 1000 K/T in Fig. 1 can be noted kinks. The nature of kinks one can to explain by changes in the activation energy, which, in turn, indicates a change in the types of carriers of electricity.

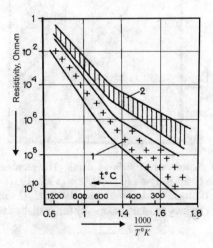

Fig. 1. Averaged diagram of specific electrical conductivity dependence of dry rocks on temperature (Parkhomenko 1965): 1 - acidic rocks (granites), 2 - basic and ultrabasic rocks (peridotites, pyroxenites)

A "K" type model with a powerful poorly conducting intermediate layer (Keller 1966, Semenov 1978, Feldman and Zhamaletdinov 2009) usually describes the picture of a gradient change in electrical conductivity with depth depending on porosity, humidity and temperature, based on laboratory data.

Estimates of temperature at great depths (deeper than 10–15 km) usually one can to carry out by extrapolating geothermal observations in wells and based on a priori data. Usually, one do it taking into account indirect estimates based on the study of metamorphic transformations in minerals, the connection of heat generation with age and composition of rocks, the involvement of chemical geothermometers and different assumptions about the petrological composition of the earth's interior (Smirnov 1968; Chermak and Ribach 1982; Glaznev 2003). Depending on the nature of the indirect information involved, the data of different authors differ significantly. As an example of such constructions, Fig. 2 shows the temperature distribution models of various authors (Clark and Ringwood 1964; Valle 1951; Semenov 1978; Shafanda et al. 1978; Cermak and Lastovichkova 1987; Zhamaletdinov 1982; Glaznev et al. 1987), characterizing the upper and lower limits of the possible temperature distribution in the lithosphere of stable regions.

One can to see that the discrepancies between the temperature estimates at a depth of 150 km reach 600 °C (from 600 to 1400 °C). In Fig. 2 shows the curves of changes

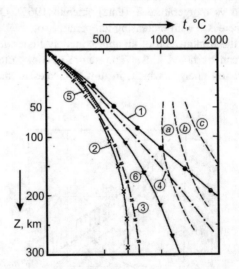

Fig. 2. The summary of geothermal profiles vs depth. The numbers on the curves indicate different models: 1 - (Clarc and Ringwood 1964), 2 - (Valle 1951), 3 - (Semenov 1978), 4- (Shafanda et al. 1978), 5 - (Cermak and Lastovichkova 1987), 6 - (Zhamaletdinov 1982; Glaznev et al. 1987). Dashed lines - melting point of rocks - dry (*a*), partially saturated with fluids (*b*) and fully saturated (*c*)

in the melting temperature with depth, calculated for rocks that are to varying degrees saturated with water (Shafanda et al. 1987). If we take the temperature distribution for

Clark and Ringwood, the melting of rocks can be observed at a depth of 130 km (at the full saturation with water) and at a depth of 160 km (at the partial saturation with water). In the case of a dry mantle, melting is absent in all of the temperature distribution models considered.

Estimates that are more detailed one can to make based on calculations of the heat balance equation, taking into account various assumptions about the distribution of energy sources in the Earth and about the physical properties of its individual layers. After the introduction of a number of simplifications (disregard by convection, by curvature of the Earth, by the time scale etc.) the heat balance equation has view (4):

$$\partial Q(z)/\partial z = q(z) \tag{4}$$

where $q(z)$ - the heat generation coefficient in units of [$\mu W/m^3$], $Q(z)$ - is the heat flux density [mW/m^2] at a depth z.

$$Q(z) = \lambda(z) \cdot G(z), \tag{5}$$

here $\lambda(z)$ - is the thermal conductivity [W/(m °C)] and $G(z) = \frac{\partial T(z)}{\partial z}$ - is the geothermal gradient [°C/m].

Methods for calculating the temperature in the Earth based on the heat balance equation differ in the ways of setting the a-priori information about the heat generation and thermal conductivity of rocks. In the paper (Shafanda et al. 1978) a layered one-dimensional model is used for this purpose, in which the radiogenic heat generation associated with the average surface heat flux changing by exponential law up to the depth of 120 km. The mantle radioactivity is accepted as constant at more depth. One can to calculate effective thermal conductivity according to a linear law, taking into account its growth with depth. The model calculated in this way for areas with a heat flux of 40 mW/m² (average for the Baltic Shield) is shown in Fig. 2a (curve 4). In the work of Glaznev et al. (1987) proposed a method for calculating heat generation at depth using seismic data. The approach is based on the use of a correlation between the decrease in the calorific value of rocks with depth and an increase in the speed of passage of longitudinal waves Vp (z) relative to mean speed V m for the upper part of the crust. In Fig. 2a shows the curve of temperature variation with depth for the Baltic Shield (curve 6), calculated based on the described approach (Glaznev et al. 1987). The thermal conductivity of the rocks of the upper part of the crust is taken within (1.5–3) W/(m °C). There is a decrease in thermal conductivity with increasing temperature (10% per 100 °C) and increasing it with increasing basicity (3.5 W/(m °C) at a depth of 300 km).

3 Technique for Estimating of the Temperature Dependence vs Depth from Results of the Deep Electrical Soundings

The original technique of estimation of temperature dependence vs depth one can find in (Spichak and Zakharova 2012). They use for this purpose the artificial neural network (ANN). The technique needs in calibration of ANN for to find the correspondence between the electrical conductivity/resistivity profiles and the data of temperature logs from the adjacent wells. But in our case the use of this method is impossible - wells are absent, the research depth is too large. By the reason we estimate temperature dependence vs depth with the use of misfit minimization between conductivity profile given by electromagnetic soundings and conductivity profile given on the base of laboratory study of rocks conductivity at high temperature and assumed rocks composition at the depth.

We will use the conventional three-layered model of the Earth's lithosphere for the Fennoscandia shield. Let we suppose that the upper part of the crust above 15 km hold water (fluids) and rocks resistivity is controlled by fluids conductivity in accordance with Archie law. Let us assume that at the upper 20 km we have the "granite" layer, consisting of acidic rocks. Lower, between the Conrad and Mohorovichić boundaries (from 20 to 40 km depth range) we have the "basalt" layer, composed by basic rocks. Following by Fig. 1 and Eq. (3), we assume that each point on the plane (log (σ), 1000/T) belongs to a certain broken line defined by two linear algebraic Eqs. (6):

$$\begin{cases} \log(\sigma) - \log(\sigma_{01}) = -0.434 \frac{E_1}{k_B \alpha} \left(\frac{\alpha}{T}\right), & T > T_{12} \\ \log(\sigma) - \log(\sigma_{02}) = -0.434 \frac{E_2}{k_B \alpha} \left(\frac{\alpha}{T}\right), & T \leq T_{12} \end{cases} \tag{6}$$

Here k_B is the Boltzmann constant ($k_B = 1.38 \cdot 10^{-23}$ J/K), $\alpha = 1000$ K is the scale constant. E_1, E_2 are values of the activation energy of different ions carrying current (in electron volts, 1 eV = $1.602 \cdot 10^{-19}$ J). T is the absolute temperature (K) on the Kelvin scale, σ is value of electrical conductivity of the rocks, σ_{01}, σ_{02} are the pre exponential coefficients for different ions that are numerically equal to the specific conductance of the rock as $T \rightarrow \infty$, T_{12} is a temperature of change of activation energy.

From (6), difference between equations is

$$\log(\sigma_{02}) - \log(\sigma_{01}) = 0.434 \frac{E_2 - E_1}{k_B \alpha} \left(\frac{\alpha}{T_{12}}\right) \tag{7}$$

It is equation for temperature of change of activation energy T_{12}.

We assume that between the boundaries broken lines for each type of rocks, each point belongs to a certain broken line with its own parameters. All broken lines between the boundaries for this type of rocks we will considered as non-intersecting. That is, we assume that all of them can to characterize by activation energy as function that vary slowly with temperature and conductivity: $E_1(T, \sigma) \approx const$, $E_2(T, \sigma) \approx const$. In this case, for given values T and σ these broken lines differ only in coefficients − σ_{01}, σ_{02}. From this it follows that for each value of the depth z, characterized by the values of the electrical conductivity of the rock and its temperature, one can enter three independent parameters σ_{01}, E_1 and T_{12}. That is, for a given depth z, we assume a correspondence of the values of temperature T and conductivity σ and the parameters σ_{01}, E_1, T_{12}. Finally, the temperature T at a given depth z, for a given type of rocks, is determined by a piecewise continuous function of log (σ):

$$T_\sigma = \begin{cases} -0.434 \frac{E_1}{k_B \alpha} \left(\frac{\alpha}{\log(\sigma) - \log(\sigma_{01})} \right), & T_\sigma > T_{12} \\ -0.434 \frac{E_2}{k_B \alpha} \left(\frac{\alpha}{\log(\sigma) - \log(\sigma_{02})} \right), & T_\sigma \leq T_{12} \end{cases} \tag{8}$$

where E_2, $\log(\sigma_{02})$ related (7) with values $\log(\sigma_{01})$, E_1 and T_{12}.

Now we will to find parameters of the earth crust model by inversion electromagnetic data, for example, jointed models electro conductivity sections for magnetotelluric (MT) and controlled source audio-magnetotelluric (CSAMT) data with supposed lithology and laboratory data for rocks conductivity. For this goal, we use the misfit functional

$$S(\sigma) = \sum_{j=1}^{N} \left(\log \rho_{\omega j}^{0}(\sigma) - \log \rho_{\omega j}^{1} \right)^2$$
$$+ \sum_{j=1}^{N} \left(\arg \left(Z_{\omega j}^{0}(\sigma) \right) - \arg \left(Z_{\omega j}^{1} \right) \right)^2 \tag{9}$$

There is $\rho_{\omega j}^{0}$ – values of apparent resistivity on frequency ω_j calculated for model of distribution of the electrical conductivity $\sigma(z)$ vs depth z. Than using (8) for known from laboratory data parameters σ_{01}, σ_{02} and E_1, E_2 for even layer of rocks one can to calculate T_σ for given three-layer geological model.

4 The Models of Temperature vs Depth Dependance for the Fennoscandian Lithosphere

The average estimation of heat generation q_m of rocks of the upper part of the granite-gneiss layer adopted in calculations equal 2.1 $\mu W/m^3$. The extrapolation of this estimate to depth was made based on a high-speed section along the Fennolora profile (Guggisberg et al. 1983), which crosses the Fennoscandia shield along the western coast of the Gulf of Bothnia. It reached the record base for registration of waves of artificial explosions (up to 2,000 km), which made it possible to build a high-speed section to a depth of 400 km. The resulting section is characterized by a gradient-step increase in Vp with a depth of 5.5–6.0 km/s at the day surface to 8.0–8.3 km/s at the Moho border (45 km) and 9 km/s at a depth of 300 km. The heat generation calculated from seismic data at a depth of 300 km is 10^{-2} $\mu W/m^3$. According to these data, the temperature section (curve 6 presented in Fig. 2) in the upper part practically coincides with the temperature curve 5 (Cermak and Lastovickova 1987) for the crust of the Earth's shields. According to curves 5 and 6 in Fig. 2, the temperature of the substance of the lithosphere at a depth of 50 km is 450 °C, and the geothermal gradient is (6–7) ° C/km; at a depth of 300 km – 1100 °C and 2 °C/km, respectively. Temperature curves do not reach the melting region, even under the assumption that the mantle is substantially saturated with fluids. This indicates indirectly the absence of an asthenosphere layer associated with partial melting of rocks (Vanyan 1984).

Taking into account the available data on the geoelectric section and petrological data on the composition of the Earth's crust and upper mantle, as well as taking into account laboratory data on the electrical conductivity of rocks under conditions of high thermodynamic parameters (Parkhomenko and Bondarenko 1972, Chermak and Lastovitskova 1987), it is possible to model the temperature of the lithosphere. The first results of such calculations are shown below in Fig. 3.

The calculation results are temperature profiles depending on the depth z for the "«normal»" and "«anomalous»" types of geoelectric section, obtained from the results of the FENICS experiment (Zhamaletdinov et al. 2015; Shevtsov et al. 2017). The first type of section, conditionally "cold", has a transversal resistance $T = 5 \cdot 10^9$ O•m². It was recorded according to the FENICS experiment data in the eastern part of the

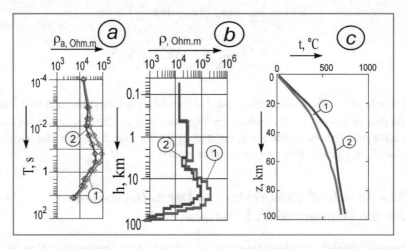

Fig. 3. The profiles example of temperature vs depth calculated for the «normal» and «anomalous» electrical sections of the Fennoscandian shield from FENICS experiment results (Zhamaletdinov et al. 2015). *a* – apparent resistivity curves for the «normal» (1) and «anomalous» (2) sections; *b* – same for resistivity vs depth profiles, *c* – same for temperature vs depth profiles

Karelian mega block and on the Kola Peninsula (Sect. 1 in Fig. 3b). The second type of Sect. (2 in Fig. 3b) conditionally "heated" is registered in the western part of the Karelian mega block and in central Finland. The second type of section can be characterized at about 70° by a higher temperature in the lower crust at the depth range 30–60 km (Fig. 3c). Thus, it can be seen that the difference in apparent resistivity of about half an order of magnitude led to a change in the temperature estimate at a depth of 30–60 km about 70 °C. At great depths up to 100 km and more the divergence of the curves again decreases to 20 °C.

5   Conclusion

In the article the algorithm is elaborated for the estimating of the temperature dependence vs depth using information given by a-priori on the change in electrical resistivity with depth (based on the results of electromagnetic soundings) and the electrical conductivity and composition of rocks at depth (based on laboratory data and geological assumptions). The solution of the problem is illustrated by the example of calculations performed for two types of geoelectrical section - «normal» and «anomalous». Information on geoelectrical sections was obtained from the results of electromagnetic soundings with industrial transmission lines (the FENICS experiment) in different parts of the Fennoscandinavian shield. The deep geological structure of lithosphere is accepted everywhere as the same. The maximum difference in apparent resistivity of about half an order of magnitude led to a change in the temperature estimate at a depth of 30–60 km about 70 °C.

Acknowledgments. This work was done with financial support of RFBR grant 18-05-00528 (rheology section) and partly with the support of the State Mission GI KSC RAS, the number of research 0226-2019-0052 (geoelectric section). The authors are deeply grateful to the lead programmer T.G. Korotkova for her help in processing of results.

References

Archie, G.E.: The electrical reistivity as an aid, determining some reservoir characteristcs. Trans. Am. Inst. Min. **146**, 54 (1942)

Brantut, N., Baud, P., Heap, M.J., Meredith, P.G.: Micromechanics of brittle creep in rocks. J. Geophys. Res. **117**, B08412 (2012). https://doi.org/10.1029/2012JB009299

Cammarano, F., Goes, S., Giardini, D., Vacher, P.: Inferring upper mantle temperatures from seismic velocities. Phys. Earth Planet. Inter. **138**, 197–222 (2003)

Chelidze, T., Chelishvily, M., Khatashvily, N., Togonidze, D., Geladze, G., Chanishvily, Z.: Electrical and magnetic properties of rocks at high temperatures and pressures, 269 p. 'Metsniereba', Tbilisi (1979)

Chermak, V., Lastovickova, M.: Temperature profiles in the earth of importance to deep electrical conductivity models, vol. 125, pp. 255–284. Pageoph (1987)

Chermak V., Ribah, L.: Thermal field of Europe, 376 p. Peace (1982)

Clark, S.P., Ringwood, A.E.: Density distribution and constitution of Mantle. Res. Geophys. **1** (1964), 35–88 (1964)

Feldman, I.S., Zhamaletdinov, A.A.: Fluid and thermal models of the conductivity of the lithosphere according to laboratory data. Complex geological and geophysical models of ancient shields. In: Proceedings of the All-Russian conference. Ed. Geological Institute KSC RAS, Apatity, pp. 100–107 (2009)

Fernandez, M., Ranalli, G.: The role of rheology in extensional basin formation modelling. Tectonophysics **282**, 129–145 (1997)

Glaznev, V.N.: Integrated geophysical models of the Fennoscandian lithosphere, 252 p. K&M, Apatity (2003)

Glaznev, V.N., Zhamaletdinov, A.A., Miroshnikov, V.P., Skopenko, G.B., Sharov, N.V.: One-dimensional geophysical model of the structure of the lithosphere of the northeast of the Baltic Shield. «normal» uts of the upper mantle (abstracts), p. 20. IG AN USSR, Kiev (1987)

Goes, S., Govers, R., Vacher, P.: Shallow mantle temperatures under Europe from P and S wave tomography. J. Geophys. Res. **105**(B5), 11153–11169 (2000)

Guggisberg, B., Ansprge, J., Mueller, S.: Structure of the uppermantle under southern Scandinavia from Fennolora data. In: First EGN Workshop, the northern segmenz, Copenhagen, 28–30 October, pp. 49–52 (1983)

Waff, H.S.: Theoretical considerations of electrical conductivity in a partially molten mantle and implications for geothermometry. J. Geophys. Res. **79**, 4003–4110 (1974)

Ioffe, A.F.: Selected issues. Mechanical and electrical properties of crystals, vol. 1, 326 p. Nauka, Leningrad (1974)

Keller, G.V.: Exploration for geothermal energy. In: Fitch, A.A. (ed.) Developments in Geophysical Exploration. Methods, vol. 2, pp. 107–150. Applied Science Publications (1981)

Keller, G.V.: Electrical properties in the deep crust. IEE Trans. Antennas Propagat. **11**(3), 344–357 (1966)

Kummerow, J., Raab, S.: Temperature dependence of electrical resistivity. Part I: experimental investigations of hydrothermal fluids. Energy Proc. **76**, 240–246 (2015)

Dai, L., Jiang, J., Li, H., Haiying, H., Hui, K.: Electrical conductivity of hydrous natural basalts at high temperatures and pressures. J. Appl. Geophys. **112**, 290–297 (2015)

Parkhomenko, E.I.: Electrical properties of rocks. Science, Moscow (1965). 164 p

Parkhomenko, E.I., Bondarenko, A.T.: Electrical conductivity of rocks at high pressures and temperatures, 279 p. Nedra, Leningrad (1972)

Quist, A.S., Marshall, W.L.: Electrical conductance of aqueous sodium chloride solutions from 0 to 800 °C and at pressures up to 4000 bars. J. Phys. Chem. **72**, 684–703 (1968)

Quist, A.S., Marshall, W.L.: Electrical conductance of some alkali metal halides in aqueous solutions from 0 to 800 °C and at pressures to 4000 bars. J. Phys. Chem. **73**, 978–985 (1969)

Semenov, A.S.: Electric section of crystalline rocks of ancient shields. Questions of Geophysics, vol. 27, pp. 108–113. Leningrad University Publishing House, Leningrad (1978)

Shafanda, J., Chermak, V., Beaudrey, L.: Methods for calculating the depth distribution of temperatures. Study of the Lithosphere Using Geophysical Methods Part 2, Kiev, pp.102–118 (1987)

Shevtsov, A.N., Zhamaletdinov, A.A., Kolobov, V.V., Barannik, M.B.: Frequency electromagnetic sounding with industrial power lines on Karelia-Kola Geotraverse. Zapiski Gornogo instituta, vol. 224, pp. 178–188 (2017). https://doi.org/10.18454/pmi.2017.2.178

Smirnov, Y.B.: The connection of the thermal field with the structure and development of the crust and upper mantle. Geotectonics **6**, 3–25 (1968)

Spichak, V., Geiermann, J., Zakharova, O., Calcagno, P., Genter, A., Schill, E.: Estimating deep temperatures in the Soultz-sous-Forêts geothermal area (France) from magnetotelluric data. Near Surf. Geophys. **13**, 397–408 (2015). https://doi.org/10.3997/1873-0604.2015014

Spies, B.R., Frischknecht, F.C.: Electromagnetic sounding. In: Nabighiam, M.N. (ed.) Electromagnetic Methods in Applied Geophysics, vol. 2, pp. 285–426. SEG (1992)

Valle, P.E.: Sull'aumento di temperature nel mantello della terra per compressions adiabatica. Ann. Geofis. **5**(4), 475–478 (1951)

Vanyan, L.L.: The electrical conductivity of the earth's crust in connection with its fluid regime. Crustal Anomalies of Electrical Conductivity, pp. 27–35. Science, Leningrad (1984)

Shafanda, Ya., Chermak, V., Bodri, L.: Methods of calculating deep temperature distributions. In: Magnitsky, V.A., Sollogub, V.B., Starostenko, V.I. (eds.) Geophysical Study of the Lithosphere (Electromagnetic and Geothermal Methods and the Joint Interpretation of Their Results. Naukova Dumka, Kiev, (1987). (in Russian)

Spichak, V.V., Zakharova, O.K.: Techniques used for estimating the temperature of the earth's interior, electromagnetic geothermometry, pp. 37–55 (2015). https://doi.org/10.1016/b978-0-12-802210-8.00002-2

Zhamaletdinov, A.A., Shevtsov, A.N., Velikhov, E.P., Skorokhodov, A.A., Kolesnikov, V.E., Korotkova, T.G., Ryazantsev, P.A., Efimov, B.V., Kolobov, V.V., Barannik, M.B., Prokopchuk, P.I., Selivanov, V.N., Kopytenko, Y.A., Kopytenko, E.A., Ismagilov, V.S., Petrishchev, M.S., Sergushin, P.A., Tereshchenko, P.E., Samsonov, B.V., Birulya, M.A., Smirnov, M.Y., Korja, T., Yampolski, Y.M., Koloskov, A.V., Baru, N.A., Poljakov, S.V., Shchennikov, A.V., Druzhin, G.I., Jozwiak, W., Reda, J., Shchors, Y.G.: Study of interaction of ELF–ULF range (0.1–200 Hz) electromagnetic waves with the earth's crust and the ionosphere in the field of industrial power transmission lines (FENICS Experiment). Proc. Atmos. Oceanic Phys. **51**(8), 826–857 (2015)

Zhamaletdinov, A.A.: «Normal» geoelectric section of the crystalline basement and its geothermal interpretation according to MHD sounding data on the Kola Peninsula. Depth electromagnetic sounding using pulsed MHD generators, pp. 35–46. Ed. KFAN SSSR, Apatity (1982)

Zharikov, A.V., Vitovtova, V.M., Shmonov, V.M., Grafchikov, A.A.: Permeability of the rocks from the Kola superdeep borehole at high temperature and pressure: implication to fluid dynamics in the continental crust. Tectonophysics **370**, 177–191 (2003)

Zhdanov, M.S., Keller, G.V.: The Geoelectrical Methods in Geophysical Exploration, p. 873 p. Elsevier, Amsterdam (1994)

On the Rheological and Geoelectrical Properties of the Earth's Crust

A. N. Shevtsov[1(✉)] and A. A. Zhamaletdinov[1,2]

[1] Kola Science Center, Geological Institute, Apatity, Russia
anshev2009-01@rambler.ru, abd.zham@mail.ru
[2] St. Petersburg Brunch of IZMIRAN, St. Petersburg, Russia

Abstract. In the present work, on the base of theoretical calculations and experimental data the transition boundary of the earth's crust from the brittle to ductile state (BDT boundary) has been studied. The results of theoretical calculations show that the position of the BDT boundary varies widely - from 5–10 to 30–40 km depending on those or other a priori (indirect) data. The use of experimental estimates allows clarifying the position of the BDT boundary. In this case, the assumption was taken that the BDT boundary coincides with the conditional "impenetrability" boundary, or else with the boundary of a sharp increase in the electrical resistivity of rocks, establishing at the depths of 10–15 km based on the results of electromagnetic soundings on direct current. Results of the deep MHD-sounding "Khibiny" and the deep DC soundings on the Murmansky block from car generator and on the Russian platform from the Volgograd-Donbass DC power transmission line were taken as the main source of experimental information.

Keywords: Brittle · Ductile · Rheology · Geoelectrics

1 Introduction

Extensive scientific literature is devoted to the consideration of the rheological parameters of the lithosphere. (Burov and Watts 2006, Bürgmann and Dresen 2008). It is believed that the tectonic structure of the geological structures observed near the surface is associated with deep processes occurring under conditions of high pressures and temperatures (Fernandez and Ranalli 1997, Goetze and Evans 1979, Jackson et al. 2008). It is believed that the division of the earth's crust into the upper and lower parts is one of the main indicators of the rheological stratification of the lithosphere (Duba et al. 2013, Dragoni 1993). It should be noted that the first information about the brittle state of the Earth's crust was received by Academician M.A. Sadovsky in studying the effects of large industrial explosions (Sadovsky 1945). The state of crust, he defined as "lumpiness". Under this term he means the fractal ordering of the elements of heterogeneity in the crust in the form of self-similar structures. The idea of MA Sadovsky was developed later in the works of Gzovsky (1975), Nikolaevsky (1986), Kirby (1983) and Ranalli (1997). Rheological stratification of lithosphere into brittle and plastic parts with the selection of the BDT (Brittle-Ductile Transition boundary) boundary has been studied most tharely in the works (Moisio and Kaikkonen 2001,

© Springer Nature Switzerland AG 2019
A. A. Zhamaletdinov and Y. L. Rebetsky (Eds.): SPS 2018, SPEES, pp. 54–61, 2019.
https://doi.org/10.1007/978-3-030-35906-5_8

Moisio 2005) on the example of the Fennoscandinavian shield. Presented paper is a continuation of this research. It consists of two sections. The first section is devoted to theoretical consideration of parameters of rheological stratification of the earth's crust. Theoretical calculations were performed using a priori information about seismicity, material composition and geothermal conditions of the earth's interior, viscosity and heat generation of rocks, stress state and strain rate, fluid mode and many other factors studied indirectly, mainly based on laboratory studies of rock samples. The second part of the work is devoted to a review of geoelectric data in order to obtain a quantitative estimate of the position of the transition boundary between the brittle and plastic states of the earth's crust (BDT boundary). At this case, it was assumed that the BDT boundary should coincide with the hypothetical boundary of "impenetrability", the boundary of a sharp increase in the electrical resistivity of rocks, installed at depths of 10–15 km on the base of DC electromagnetic sounding data. Results of the deep MHD-sounding "Khibiny" (Velikhov 1989, Zhamaletdinov 1990) and the deep DC soundings on the Murmansky block from car generator (Kolobov et al. 2018) and on the Russian platform from the Volgograd-Donbass DC power transmission line (Zhamaletdinov et al. 1982) were taken as the main source of experimental information.

2 Rheology

The rheology of rocks establishes relationships between the properties of rocks and parameters characterizing their state. The typical rheological properties of rocks include elasticity, creep, relaxation, elastic consequence, fluidity et al. In fact, rheological equations do not describe a real rock, but its model is an ideal body (material). The number of rheological models is not limited; however, all of them one can to represent by a combination of three simple bodies (models) - Hooke's body (elasticity), Newton's body (viscosity), Saint-Venant's body (plasticity). Mathematically, these models one can to describe by equations linking the components of the strain tensors and their velocities with the components of the stress tensors and their derivatives with respect to time.

For simplicity, we confine ourselves to the representation of the lithosphere in the form of several rigid shells with viscoelastic properties, plasticity and creep depending on temperature. Following (Ranalli 1997, Moisio 2005), we will consider the lithosphere as a set of solid shells obeying first-order equations.

$$R\left(\sigma, \frac{d\sigma}{dt}, \varepsilon, \frac{d\varepsilon}{dt}, \{M_i\}, \{S_i\}\right) = 0. \tag{1}$$

Equations (1) relate the deformation ε and its rate $\frac{d\varepsilon}{dt}$ to the stress σ and its rate $\frac{d\sigma}{dt}$, and also include the material characteristics $\{M_i\}$ (elastic characteristics, viscosity etc.) and state parameters $\{S_i\}$ (temperature, pressure, etc.), $i = 1, 2, \ldots, N$ – index of shells.

In fragile shells, for existing systems of fault planes and the possibility of rock sliding along these planes, an approximate Amonton's law is assumed: the tangential stresses on the contact surface S_τ are proportional to normal stresses. Amonton's law is supplemented by the Navier-Coulomb criterion: normal stress, acting in the plane of shear failure, increases the body's resistivity to shear by an amount proportional to the magnitude of this normal stress. The destruction of a solid body in this case will occur when the shear stress acting in the shear plane reaches a value determined by the expression (2).

$$\tau = \tau_0 + \mu \cdot \sigma \tag{2}$$

Here τ_0 is the Coulomb shear strength, μ is the coefficient of internal friction. In porous rocks saturated with non-drained fluid, the Navier – Coulomb criterion has a similar appearance.

$$\tau = \tau_0 + \mu \cdot \sigma_e \tag{3}$$

Where the effective stress in Eq. (4) reflects an increase in shear instability with an increase in the pore pressure P_n

$$\sigma_e = \sigma_a - P_n. \tag{4}$$

This reduces the total shear resistivity. Here, the stress σ_a, acts between the solid components of the rock.

The Byerlee's law (Byerlee 1967), used for the upper – fragile part of the crust, is a combination of the Amonton law with the Navier – Coulomb criterion (Ranalli 1997, Moisio 2005). Explicitly in terms of the differential stress (the difference between the maximum σ_1 and minimum σ_3 values of the main stress (Pa)), lithostatiç pressure $P_\lambda = \rho g z$ and pore pressure P_n, the Byerlee's law is written as (Moisio 2005)

$$\sigma_1 - \sigma_3 = \alpha \rho g z (1 - \lambda), \tag{5}$$

$$\alpha = \begin{cases} \frac{r-1}{r}, & (a) \\ r - 1, & (b) \\ \frac{r-1}{1+\beta\cdot(r-1)}, & (c) \end{cases} \tag{6}$$

In (6) we have (a) – across to fault, (b) – along to fault, and (c) – shift and fault. Here parameter r is

$$r = \left(\sqrt{1+\mu^2} - \mu\right)^{-2} \tag{7}$$

Here $\lambda = P_n/P_\lambda$ is the pore factor - the ratio of pore pressure to the lithostatic one, ρ is the density of the rock (kg/m^3), z is the depth (m), g is the acceleration of gravity (m/s^2), μ is the coefficient of sliding friction, $\beta = \frac{\sigma_2 - \sigma_3}{\sigma_1 - \sigma_3}$ - is the relative difference of the intermediate stress σ_2. These relations relate only to the case of sliding along the already existing faults in the brittle shell.

The plastic deformation of silicate polycrystals is usually caused by dislocation creep, with the exception of very fine-grained (<10 μm in order of magnitude) rocks. Dislocation creep is a consequence of the motion of dislocations in the structure of the crystal lattice. This mechanism is temperature dependent. At high stresses, dislocation creep is often dominant, although diffusion creep can also affect the overall strain rate.

The dislocation creep rate $\dot{\varepsilon} = \frac{d\varepsilon}{dt}$ has an exponential dependence on temperature and pressure and a nonlinear relationship with the stress in the power form. The dislocation motion one can to consider equivalent to a non-Newtonian flow if the required activation threshold is exceeded. Then the basic equation one can to write using the empirical power law (8) according to (Kirby 1983)

$$\sigma_1 - \sigma_3 = \left(\dot{\varepsilon}/A_p\right)^{1/n}\exp\left(E_p/nRT\right) \tag{8}$$

Here $\dot{\varepsilon} = \frac{d\varepsilon}{dt}$ (c^{-1}) – strain rate, T – temperature (K), R = 8.314 J mole^{-1} K^{-1} – universal gas constant, A_p (Pa^{-n}/s), n, E_p (J mol^{-1}) – empirically determined rock parameters (Table 1).

Table 1. Creep parameters in some rocks and minerals of the lithosphere (Ranalli 1995)

Material	A_p (MPa^{-n}/s)	n	E_p (J mol $^{-1}$)
Granite	1.8×10^{-9}	3.2	123
Granite (wet)	2.0×10^{-4}	1.9	137
Quartz diorite	1.3×10^{-3}	2.4	219
Felsic granulite	8.0×10^{-3}	3.1	243
Peridotite (dry)	2.5×10^{4}	.5	532

At high pressures ($P > 200$ MPa), the power law is violated, and other approximations are used. So, for example, for the olivine mineralogy for typical mantle conditions, Dorn's law (8) is applied (Goetze and Evans 1979)

$$\sigma_1 - \sigma_3 = \sigma_D\left(1 - \left(-RT/E_D \ln(\dot{\varepsilon}/A_D)\right)^{1/2}\right) \tag{9}$$

Here s_D (Pa), A_D (s^{-1}), n, E_D (J mol $^{-1}$) are empirically determined rock parameters.

Typically, the strain rate one can to consider a constant for all hard shells and it is 10^{-16} - for "slow" and 10^{-14} - for "fast" processes.

For example, Fig. 1 shows the results of the calculation of the parameters of the transition from the brittle to ductile state (the BDT boundary, as defined by Moisio 2005). This is results of the calculations for simplified model presented in Table 2.

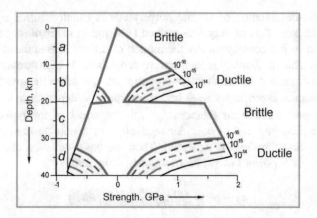

Fig. 1. Rheological profiles for compression deformation $\alpha = -0.75$ ($\Delta\sigma = \sigma_1 - \sigma_1 < 0$ and stretching $\alpha = 3$ ($\Delta\sigma > 0$), for different strain rates from "slow" $\dot{\varepsilon} = 10^{-16}\ \text{s}^{-1}$ - fat solid line, to "fast" $\dot{\varepsilon} = 10^{-14}\ \text{s}^{-1}$ - thin solid line with a step of $10^{0.5}$, indexes of the curves is $\dot{\varepsilon}\ (\text{s}^{-1})$. The designations in column: a - wet granites; b - dry granites; c - quartz-diorite; d - granulite

Table 2. Model for the calculation of depth profiles of rheological profiles. ρ, kg/m³ - density, k_Π, %, porosity

Layer	Composition	Thickness (km)	ρ, kg/m³	k_Π, %
Fractured upper crust	Wet granite	.5	2560	3
Wet upper crust	Wet granite	2.5	2600	1
Wet middle crust	Wet granite	7	2700	0.5
Middle crust - dry	Dry granite	10	2700	0.1
Low crust - dry	Quartz-diorite	10	2850	0.05
	Felsic granulite	10	2900	0.01
Mantle	Dry peridotite	∞	3200	0.01

From results of calculations presented in the Fig. 1, one can to see that location of the BDT transition boundary can vary in the Fennoscandian shield in the widest limits (from 12–17- to 30–40 km), depending on a priori assumptions. The composition of rocks, temperature and characteristic time $t = 1/\dot{\varepsilon}$ play a decisive role in the position of the BDT boundary. "Slow" deformations characterize processes lasting billions of years $\left(\dot{\varepsilon} = 10^{-16}\ \text{s}^{-1}\right)$, "fast" - millions of years $\left(\dot{\varepsilon} = 10^{-14}\ \text{s}^{-1}\right)$.

3   Geoelectrics

At the quantitative estimates of the nature of geophysical boundaries in the crystalline crust, certain expectations were associated with geoelectromagnetic methods, in particular, with magnetotelluric soundings (MTS). It was assumed "a- priori" that geophysical boundaries should be accompanied by a sharp increase in electrical conductivity in the form of intermediate conductive layers, the nature of which may be associated with temperature gradients, changes in the chemical composition of rocks or fluid saturation of the lithosphere (Vanyan 1965). However, reliable solutions have so far not been obtained due to the multi-factorial interpretation of results obtained and due to the ambiguity of inverse problem solution of induction electromagnetic soundings under conditions of high resistive media.

In this article the possibility of existence of a geophysical boundary of the opposite property in the continental crust, in the form of a high resistive basement at a depth of 10–15 km is substantiated. This boundary marks the transition zone between the upper, fragile part of the earth's crust ("brittle zone") and the lower, plastic part ("ductile zone"). It is defined by us as the "impenetrability" boundary for galvanic (DC) currents. During induction sounding under the conditions of the far (wave) zone, in particular, with magnetotelluric sounding (MTS), layers with high resistivity, if their thickness is less than the length of electromagnetic wave in the Earth's crust, fall into the area of "transparency" and their detection becomes problematic.

The first experimental indication of existence of extremely high resistive (conditionally "impenetrable" for DC current) basement at the depth of 10–15 km have been obtained during the deep electromagnetic sounding with the use of impulsive MHD generator "Khibiny" of 80 MW (Velikhov et al. 1989). According to these data, it has been established that ap-proximately 20% of MHD generator's current penetrates into the Earth's crust by galvanic way and propagates inside the upper layer of about 10 km thickness. Its distribution obeys the law of logarithmic potential, i.e. current density (electric field intensity) decreases in proportion to the first power of the distance between transmitter and receiver. This result has been interpreted at the first step in frame of J. Söderholm's hypothesis about two-layered structure of Precambrian crystalline crust (Zhamaletdinov 1990). The upper, "proto-sedimentary" layer has an integral longitudinal conductivity of 1 S and an average resistivity of 10^4 Ω-m. It is composed by primary sedimentary metamorphic complexes and volcanogenic rocks of the Proterozoic and Upper Archaean age. Within the upper layer so-called crustal electronically conductive structures are widespread, they are represented by sulfide-carbonaceous rocks of organic origin. Their extension to the depth does not exceed 10 km. This result was obtained by different methods - by separation of EM field of MHD generator "Khibiny" on the galvanic and induction modes (Zhamaletdinov 1990, Kirillov and Osipenko 1984) by MHD signals processing with the use of migration method (Zhdanov and Frenkel 1983) and by processing of magnetovariational profiling results (Zhamaletdinov et al. 1980). Below the inhomogeneous proto-sedimentary upper layer of moderate high resistivity (10^4 Ω m) a very high resistive resistivity of 10^5 Ω m proto-basement is situated with and more 10–15 km thickness and. It is believed that the proto-basement has a low Archaean age and is formed at the early nuclear stage of the Earth's development.

Subsequently, the described above "geological" interpretation of two-layered model of the Earth's crust electrical conductivity, followed from of MHD "Khibiny" experiment, and were abandoned in favor of the physical model which suggests its subdivision into a brittle part and ductile part (Zhamaletdinov 2011). However, the elements of continuity between the two models described are preserved.

The existence of extremely high resistive primary basement in the view of an "impenetrability" boundary in the continental crust was confirmed by results of "Volga" experiment on the deep DC sounding in the field of industrial power line "Volgograd-Donbass" of 470 km length (Zhamaletdinov et al. 1982). According to results of "Volga" experiment the ascending asymptotic of apparent resistivity that has been followed up to distance of 700 km between transmitter and receiver. This indicates a high transversal resistance of the lithosphere, an order of about $7 \cdot 10^9 \ \Omega \ m^2$ and the absence of the effect of direct current penetration into the lower half-space due to existence of "impenetrability" boundary at the depth below 10 km (Zhamaletdinov et al. 2011).

Acknowledgments. This work was done with financial support of RFBR grant 18-05-00528 (rheology section) and partly with the support of the State Mission GI KSC RAS, the number of research 0226-2019-0052 (geoelectric section). The authors are deeply grateful to the lead programmer T. G. Korotkova for her help in processing of results.

References

Burov, E.B., Watts, A.B.: The long-term strength of continental lithosphere: 'jelly sandwich' or 'creme brulee'? GSA Today **16**, 4–10 (2006)

Byerlee, J.: Friction characteristics of granite under high confining pressure. J. Geophys. Res. **72** (14), 3639–3648 (1967). https://doi.org/10.1029/JZ072i014p03639

Dragoni, M.: The brittle-ductile transition in tectonic boundary zones. Ann. Geophys. Annali di Geofisica **36**(2), 37–44 (1993). https://doi.org/10.4401/ag-4282

Duba, A.G., Durham, W.B., Handin, J.W., Wang, H.F. (eds.). The Brittle-Ductile Transition in rocks. Geophysical Monograph Series, vol. 56, p. 243. AGU (2013) https://doi.org/10.1029/gm056

Fernandez, M., Ranalli, G.: The role of rheology in extensional basin formation modelling. Tectonophysics **282**, 129–145 (1997)

Goetze, C., Evans, B.: Stress and temperature in the bending lithosphere as constrained by experimental rock mechanics. Geophys. J. R. Astr. Soc. **59**, 463–478 (1979)

Gzovsky, M.V.: Basics of Tectonophysics, p. 535. Science, Moscow (1975)

Jackson, J., McKenzie, D., Priestley, K., Emmerson, B.: New views on the structure and rheology of the lithosphere. J. Geol. Soc. **165**(2008), 453–465 (2008)

Kirby, S.H.: Rheology of the lithosphere. Rev. Geophys. **21**, 1458–1487 (1983)

Kirillov, S.K., Osipenko, L.G.: Study of the Imandra-Varzuga conductive zone (Kola Peninsula) with the use of MHD generator. In: Crustal Anomalies of Electrical Conductivity, pp. 79–86. Nauka, Moscow (1984). (in Russian)

Kolobov, V.V., Barannik, M.B., Ivonin, V.V., Selivanov, V.N., Zhamaletdinov, A.A., Shevtsov, A.N., Skorokhodov, A.A.: Experience of using the generator "Energy-4" for remote and frequency electromagnetic soundings in the experiment "Murman 2018". In: Proceedings of the Kola Science Center of RAS, 2018, №17, pp. 7–20 (2018). https://doi.org/10.25702/ksc.2307-5252.2018.9.8.7-20

Moisio, K., Kaikkonen, P.: Geodynamics and rheology along the DSS profile SVEKA'81 in the central Fennoscandian Shield. Tectonophysics **340**, 61–77 (2001)

Moisio, K.: Numerical lithospheric modelling rheology stress and deformation in the Centrak Fennoscandian Shield. Academic dissertation, p. 39. Univ. of Oulu (2005)

Nikolaevsky, V.N.: Dilatance rheology of lithosphere and tectonic stress waves. The main problems of seismotectonics, pp. 51–68. Science, Moscow (1986)

Bürgmann, R., Dresen, G.: Rheology of the lower crust and upper mantle: evidence from rock mechanics, geodesy, and field observations. Ann. Rev. Earth Planet. Sci. **2008**(36), 531–567 (2008). https://doi.org/10.1146/annurev.earth.36.031207.124326

Ranalli, G.: Rheology of the Earth, 2nd edn., xv 413 p. Chapman & Hall, London, Glasgow, Weinheim, New York, Tokyo, Melbourne, Madras (1995)

Ranalli, G.: Rheology and deep tectonics. Annali di Geofisica **XL**(3), 671–680 (1997)

Sadovsky, M.A.: Experimental studies of the mechanical action of a shock wave of an explosion. In: M.-L. 1945 Proceedings of the Seismological Institute of the Academy of Sciences of the USSR, no. 11, pp. 125–146 (1945)

Vanyan, L.L.: Principles of Electromagnetic Soundings, p. 107. Nedra, Moscow (1965)

Velikhov, E.P. (ed.) Geoelectrical Investigations with Powerful Source of Current on the Baltic Shield, p. 272. Nauka, Moscow (1989). (In Russian)

Zhamaletdinov, A.A.: Model of Electrical Conductivity of Lithosphere by Results of Studies with Controlled Sources (Baltic Shield, Russian Platform), p. 159. "Nauka", Leningrad (1990). (in Russian)

Zhamaletdinov, A.A.: The new data on the structure of the continental crust based on the results of electromagnetic sounding with the use of powerful controlled sources. Doklady Earth Sciences 2011, vol. 438, Part 2, pp. 798–802 (2011). ISSN 1028_334X

Zhamaletdinov, A.A., Kovalevsky, V.Ya., Pavlovsky, V.I., Tanachev, G.S., Tokarev, A.D.: Deep electrical sounding with DC power line of 800 kV "Volgograd - Donbass", USSR 1982. Doklady Earth Sciences, vol. 265, no 5, pp. 1101–1105 (1982)

Zhamaletdinov, A.A., Kulik, S.N., Pavlovsky, V.I., Rokityansky, I.I., Tanatchev, G.S.: Anomaly of short-period geomagnetic variations over the Imandra-Varzuga structure (Kola Peninsula). Geofiz. J. **2**(1), 91–96 (1980). (in Russian)

Zhamaletdinov, A.A., Shevtsov, A.N., Korotkova, T.G., et al.: Deep electromagnetic sounding of the lithosphere in the Eastern Baltic (Fennoscandian) shield with High_Power controlled sources and industrial power transmission lines (FENICS Experiment). Izvestiya. Phys. Solid Earth **47**(1), 2–22 (2011)

Zhdanov, M.S., Frenkel, M.A.: Migration of electromagnetic fields in solving inverse problems of geoelectrics. DAN USSR. Doklady Earth Sciences, vol. 271, no. 3, pp. 589–594 (1983)

Experimental Study of Lithosphere Structure

Experimental Study of Impermeability Boundary in the Earth Crust

A. A. Zhamaletdinov[1,2(✉)], A. N. Shevtsov[1], A. A. Skorokhodov[1],
V. V. Kolobov[3], and V. V. Ivonin[3]

[1] Kola Science Center, Geological Institute, Apatity, Russia
abd.zham@mail.ru, anshev2009-01@rambler.ru
[2] Saint-Petersburg Branch of IZMIRAN, Saint Petersburg, Russia
[3] NERC KSC RAS, Apatity, Russia
l_i@mail.ru

Abstract. Results of study of supposed boundary of impenetrability in the crystalline earth crust are discussed in the article. The Murman-2018 experiment on distance DC sounding in combination with frequency and audio magnetotelluric sounding was undertaken to solve the problem. Methods of field work and data processing are presented in the first part of the work. The second part of the work is devoted to the interpretation method description and to presentation of results obtained. Impenetrability boundary is detected in the shape of the sharp resistivity increase from 10^4 to 10^6 $\Omega \cdot$ m at a depth of 10–15 km. That boundary is accepted as transition zone between the brittle and ductile states of the earth crust substance (Brittle-Ductile Transition zone, BDT). To obtain the more reliable information on parameters of the impenetrability boundary it is necessary to conduct additional DC studies using remote sensing at distances up to 1000 km between transmitter and receiver.

Keywords: Murmansky block · Distance DC sounding · Impenetrability boundary · Brittle state · Ductile state · Earth crust

1 Introduction

Results of electromagnetic soundings of the Baltic Shield lithosphere with the use of EM field powerful controlled sources allowed to make proposition that the crystalline crust consists of two strata separated between each other by a sharp increase of resistivity at a depth of about 10–15 km (Zhamaletdinov 1990). The upper crust has a lower electrical resistivity (around 10^4 $\Omega \cdot$ m) and a sharp horizontal heterogeneity. The lower part, on the contrary, is distinguished by abnormally high resistivity (10^5–10^6 $\Omega \cdot$ m) and high horizontal homogeneity of electrical properties. Following by (Nikolaevsky 1996; Ranalli 2000; Moisio and Kaikkonen 2004), it is believed that the upper layer represents brittle crust, and the lower part is ductile or, rather, quasi-ductile (Zhamaletdinov 2011). The boundary of resistivity increase, conventionally called as boundary of "impenetrability", is an obstacle to the penetration of direct current to the depth and is identified with the boundary of the earth's crust substance transition from the brittle to quasi-ductile state (BDT zone) (Moisio 2005) and with a seismic Conrad boundary (Zhamaletdinov 2014). However, so far no quantitative estimates have been

© Springer Nature Switzerland AG 2019
A. A. Zhamaletdinov and Y. L. Rebetsky (Eds.): SPS 2018, SPEES, pp. 65–71, 2019.
https://doi.org/10.1007/978-3-030-35906-5_9

obtained of the position of the "impenetrability" boundary. This is explained by the fact that studies of the deep electrical conductivity of the Earth are carried out mainly with the use of induction, magnetotelluric soundings. At the same time, it is known that during induction soundings, layers of high resistivity, the thickness of which is less than electromagnetic wave length in the ground, fall into the region of "transparency" and their detection, and moreover, the separation between them becomes problematic (Vanyan 1997). The Murman-2018 experiment on the distance sounding (DS) in conjunction with CSAMT and AMTS, was performed in order to fill the gap and investigate the supposed boundary of "impenetrability". The experimental technique and the main results obtained compile the basis of this article.

2   The Technique

The Murman-2018 experiment was carried out on the territory of the Murmansk block, representing a monotonous geological province composed by ancient granite-gneiss rocks of the lower Archaean age (Batieva et al. 1978). The Murmansk block territory is distinguished by high homogeneous electrical resistivity of around 10^4 $\Omega \cdot$ m and by almost complete absence of ore prospective objects. The width of the Murmansky block is of 50–70 km with a length of 400–500 km. In the northeast side it is bordered by the Barents Sea, in the southwest it is bounded by younger formations of Kolmozero-Voronya and Keivy structures. The scheme of Murman-2018 experiment with location of grounded transmitting dipoles AB1, AB2 and traces 1, 2, 3 of receiving sites is shown in Fig. 1.

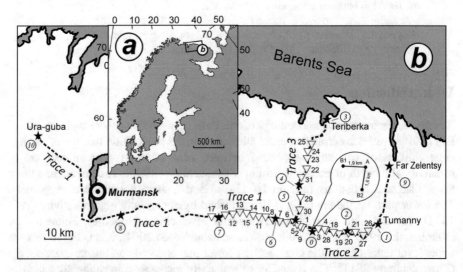

Fig. 1. The scheme of the "Murman-2018" experiment. a - contours of Fennoscandian (Baltic) Shield with the position of the sounding area, indicated by a rectangle "b"; b - area of work with the position of the feeding grounded dipoles AB1, AB2 and traces 1, 2, 3 of the deep soundings. The DS sounding sites are shown by triangles with numbers 1–31. Frequency sounding sites (CSAMT) in combination with AMTS are shown by asterisks with numbers 0–10 in circles

Soundings were carried out along three traces - in the directions to the West (trace No. 1 to Murmansk-Ura-Guba), to the East (trace No. 2 to Tumanny-Far Zelentsy) and to the North (trace No. 3 to Teriberka), (Fig. 1b). The maximum separation between the source and receiver reached 105 km. Measurements at each receiving site were carried out with the transversal (subequatorial) and longitudinal (sub axial) arrangements of the feeding grounded dipoles AB1 and AB2 relatively to sounding traces 1, 2 and 3 (Fig. 1b).

Two mutually orthogonal grounded electric dipoles with a length of 1.9 and 1.6 km (Fig. 1b) served as feeding lines. The current was supplied by turns into grounded lines AB1 and AB2 from the car generator "Energia-4" of 29 kW power (Kolobov et al. 2013). Output voltage was of 1200 V in the frequency range from 1 Hz to 2 kHz. The total active resistance of AB1 line was 55 Ω, of AB2 line – 57 Ω.

3 Distance Sounding

Distance soundings (DS) were performed at 31 sites (Fig. 1b). The step of distance changing between transmitter and receiver in the interval of 5–56 km was applied not logarithmic, as it is usually accepted, but linear (5 km) in order to detect the influence of horizontal heterogeneity on observations results. The current into the feeding grounded lines AB was supplied in the form of rectangular bipolar signals in the shape of meander with a period of T = 5.15, 2.58 or 1.06 s. The period of the current has been chosen in the receiving site depending the distance to the source and the signal-to-noise ratio. The recording was made using a VMTU-10 wideband 32-bit magnetotelluric station (Kopytenko et al. 2010). The measurements were carried out along the magnetic meridian (Ex) and in latitude (Ey) using MN = 2 × 50 m lines with a midpoint.

Signal processing was carried out by means of two methods - in the spectral mode and in the accumulation mode. Spectral processing was reduced to determining the amplitude of the first harmonic. The spectrum was determined through the power-density spectrum. The data accumulation was carried out by time series stacking for the first harmonic with the main period T and for the second harmonic with the half period of 0.5T. The amplitude ratio of the first and second harmonics has been used to estimate the signal-to-noise ratio.

The apparent electrical resistivity was calculated with the use of geometrical coefficient for the total vector of electric field intensity (Zhamaletdinov 2011) using equation:

$$\rho_k = k \cdot \frac{\sqrt{E_x^2 + E_y^2}}{I}, [O\mathcal{M} \cdot \mathcal{M}],$$

where geometric factor k is determined by expression.

$$k = 2\pi \left[r_{AO}^{-4} + r_{BO}^{-4} - \frac{r_{AO}^2 + r_{BO}^2 - r_{AB}^2}{(r_{AO} \cdot r_{BO})^3} \right]^{-\frac{1}{2}}, \left[\mathcal{M}^2 \right],$$

where r_{AO} and r_{BO} are distances to the central reception point O from groundings A and B, respectively, r_{AB} is the length of the feeding line AB.

4 Data Interpretation

In the area of supply lines AB1 an AB2 the vertical soundings VES were carried out in different directions at distances of up to 2 km for the purpose to estimate the media homogeneity at the primary field excitation place (Fig. 2).

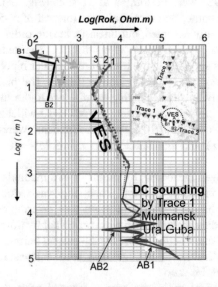

Fig. 2. Results of local vertical DC soundings VES along feeding lines AB1 and AB2 (VES 1, 2 and 3) and regional deep DC soundings along trace 1 to Murmansk and Ura-Guba

With the further increase in separation between transmitter and receiver from 2–3 to 30–40 km, the behavior of the curves of apparent resistance has changed dramatically. In the lower part of Fig. 2, one can see that the curves have acquired a saw-tooth character, which does not fit into the one-dimensional model. Values range from 4 to 70 thousand ohm meters. At the same time, instead of the expected continuation of a smooth increase in values with spacings increasing, there is some stabilization of values at the level of the average meaning of $2 \cdot 10^4 \, \Omega \cdot m$. In Fig. 2 it can be seen that the differences in values are more pronounced on the equatorial setting measured with the meridional line AB2. This fact, as well as the incoherent, sometimes antiphase behavior of the axial and equatorial sounding curves, indicates the two-dimensional nature of the section and the predominant influence of the submeridional faults and fracturing zones. The mentioned differences in values are obviously associated with the influence of sub-vertical zones of fracturing and faults, which are sloping (flattening) with depth. It can be assumed that by their influence can be explained existence of intermediate conductive layer of the dialate-diffusion nature (DD layer) at the depth of 2–7 km described in (Zhamaletdinov et al. 1998). This area of the geoelectric section in the range of distances from 10 to 30–40 km, corresponding to the interval of depths from 2–3 to 10–15 km, we called the compaction zone. In this part of the geoelectric section, the electrical conductivity of rocks is determined by the existence of free fluids in open cavities, which can be formed due to the phenomenon of dilatancy under the combined influence and opposition of lithostatic and tangential pressures (Nikolaevsky 1996; Zhamaletdinov et al. 2017). Together with the upper layer of the gradiental increase in resistivity, the upper crust (compaction zone and DD-layer) till the depth of 10–15 km can be defined as an area of the brittle state of the earth's crust.

With further increase in the distance between the source and receiver up to 70–100 km, the values quickly increase to 150–300 thousand. This indicates the existence at the depth of the base that is characterized by higher resistance. This boundary is conditionally defined as the "impenetrability" boundary, which may be associated with the transition of rocks from a brittle to a quasi-plastic (plastic) state. Thus, due to the continuous (linear) change in spacing, we managed to make a qualitative idea of the nature of the geoelectric section and present it as an "A" type section cut with a gradient - stepwise increase of resistivity with depth. A quantitative estimate of geoelectric section parameters could be obtained with the use of inversion problem solution (Fig. 3).

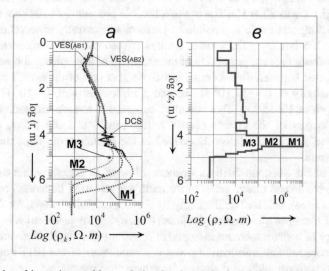

Fig. 3. Results of inversion problem solution by the method of fitting of the forward problem solutions on the example of distance DC soundings along the Murmansk-Ura-guba trace (the position of the sounding trace is shown in Fig. 1b). DCS – the averaged curve from equatorial and axial apparent resistivity curves

The fitting results reflect three possible variants of the geoelectrical section model, differing between each other by the magnitude of resistivity of the low conductive layer in the lower part of the earth crust. Its resistivity varies from $10^5 \, \Omega \cdot$ m for M1 model to $10^6 \, \Omega \cdot$ m for M2 model and to $10^7 \, \Omega \cdot$ m for M3 model. To obtain a more complete agreement with experimental data the intermediate layer of reduced resistivity ($10^4 \, \Omega \cdot$ m) is inserted on all three models at the depth interval from 2 to 7 km. It is associated with the DD layer (Zhamaletdinov et al. 2017). Due to the inserted DD layer the inflection of apparent resistivity curve in the spacing interval from 10 to 30–40 km was confirmed in theoretical calculations. It can be noted in Fig. 3 that the resolution ability of DC apparent resistivity curves concerning to parameters of a high resistivity basement is rather small. Changes in resistivity of M1–M3 models from $10^5 \, \Omega \cdot$ m to $10^7 \, \Omega \cdot$ m (three orders of magnitude) resulted in apparent resistivity changes of less than 1 order. Of the three models, preference can be given to the model M2 with a base resistivity of $10^6 \, \Omega \cdot$ m. However, the complete solution of the inverse problem in the Murman-2018 experiment requires additional studies taking into account the anisotropy of the intermediate layer of reduced resistivity and involving 2D models that take into account the flattening (sloping, bending) of faults and cracks with depth.

5 Summary

On the territory of the Murmansky block the "Murman 2018" experiment was implemented on the deep remote sensing DC sounding in combination with frequency and audio magnetotelluric soundings.

As a current source the autonomous generator "Energia-4" with a power of 29 kW and an output voltage of up to 1200 V was used, allowing sounding in the frequency range from 0.1 Hz to 2000 Hz. The use of a flexible procedure of stacking and spectral processing of signals made it possible to perform successfully observations at a distance up to 105 km from the source. Results of Murman-2018 experiment made it possible to substantiate the possibility of the existence of a geophysical boundary in the continental crust between the moderate resistive upper part of the earth's crust with an average resistivity of $(2–3) \cdot 10^4 \, \Omega \cdot$ m and a high resistive (10^6–$10^7 \, \Omega \cdot$ m) lower part at a depth of 10–15 km. This boundary can mark a transitional zone between the upper, brittle part of the earth's crust ("brittle zone") and the lower, ductile part ("ductile zone") and is conditionally defined as the boundary of "impermeability" for galvanic currents.

Results of the Murman-2018 experiment allow us to conclude that the boundary between the brittle and plastic states of the earth's crust (BDT boundary, brittle-ductile transmission zone), assessed qualitatively by rheological calculations, for the first time in the world literature quantitatively estimated according to geoelectrics in the form of the boundary of a sharp increase in apparent resistivity at a depth of about 10–15 km. To obtain more reliable information about the parameters of the BDT boundary, it is necessary to perform additional studies, by conducting remote sensing and frequency soundings of increased accuracy with the use of powerful controlled sources.

Acknowledgments. This work was supported by the grant of the RFBR 18-05-00528 (expedition) and partly with the support of the State Mission, GI KSC RAS, the subject of research 0226-2019-0052 (interpretation). The authors are deeply grateful to the lead programmer T.G. Korotkova for her help in processing and interpreting the results and V.E. Kolesnikov for his help in topographic data binding and its design.

References

Batieva, I.D., Belkov, I.V., Vetrin, V.R., Vinogradov, A.N., Vinogradova, G.V., Dubrovsky, M. I.: Granitoid Precambrian formations of the northeastern part of the Baltic Shield. Nauka, 263 p. (1978)

Kolobov, V.V., Barannik, M.B., Zhamaletdinov, A.A.: Generation and measuring complex "Energiya" for the lithosphere electromagnetic sounding and for the seismically active zones monitoring. In: SPb: SOLO 2013, 240 p. (2013)

Kopytenko, E.A., Palshin, N.A., Poljakov, S.V., Schennikov, A.V., Reznikov, B.I., Samsonov, B.V.: New portable multifunctional broadband MT System. In: IAGA WG 1.2 on Electromagnetic Induction in the Earth 20th Workshop Abstract, Egypt (2010)

Moisio, K., Kaikkonen, P.: The present day rheology, stress field and deformation along the DSS profile FENNIA in the central Fennoscandian Shield. J. Geodyn. **38**, 161–184 (2004)

Moisio, K.: Numerical lithospheric modelling rheology stress and deformation in the Centrak Fennoscandian Shield. Acad Dissert. Univ. of Oulu, 39 p. (2005)

Nikolaevsky, V.N.: Cataclastic breaking down of rocks of earth crust and anomaly of geophysical fields. Izv. Akad. Nauk, Ser. Fiz. Zemlin, no. 4, p. 41–50 (1996)

Ranalli, G.: Rheology of the crust and its role in tectonic reactivation. J. Geodyn. **30**, 3–15 (2000)

Vanyan, L.L.: Electromagnetic Soundings. Nauchny Mir, Moscow, 218 p. (1997)

Zhamaletdinov, A.A.: The new data on the structure of the continental earth crust based on the results of electromagnetic with the use of powerful controlled sources. Doklady Earth Sci. **438**, 798–802 (2011). Part 2

Zhamaletdinov, A.A.: Model of electrical conductivity of lithosphere by results of studies with controlled sources (Baltic shield, Russian platform). Leningrad. Nauka, 159 p. (1990)

Zhamaletdinov, A.A.: The nature of the Conrad discontinuity with respect to the results of kola superdeep well drilling and the data of a deep geoelectrical survey. Doklady Earth Sci. **455**, 350–354 (2014). Part 1, ISSN 1028_334X

Zhamaletdinov, A.A., Shevtsov, A.N., Tokarev, A.D., Korja, T., Pedersen, L.: Experiment on the deep frequency sounding and dc measurements in the central finland granitoid complex. In: 14th Workshop in Sinaia Electromagnetic Induction in the Earth (Romania), p. 83 (1998)

Zhamaletdinova, A.A., Velikhov, E.P., Shevtsov, A.N., Kolobov, V.V., Kolesnikov, V.E., Skorokhodov, A.A., Korotkova, T.G., Ivonin, V.V., Ryazantsev, P.A., Biruly, M.A.: The Kovdor-2015 experiment: study of the parameters of a conductive layer of dilatancy–diffusion nature (DD Layer) in the archaean crystalline basement of the Baltic Shield. Doklady Earth Sci. **474**, 641–645 (2017). Part 2

Geoelectric Models Along the Profile Crossing the Indian Craton, Himalaya and Eastern Tibet Resulted from Simultaneous MT/MV Soundings

Iv. M. Varentsov[1]([✉]), P. V. Ivanov[1], I. N. Lozovsky[1], D. Bai[2], X. Li[2], S. Kumar[3], and D. Walia[3]

[1] Geoelectromagnetic Research Center,
Schmidt Institute of Physics of the Earth RAS, Moscow, Troitsk, Russia
`ivan_varentsov@mail.ru`
[2] Institute of Geology and Geophysics, CAS, Beijing,
People's Republic of China
[3] North-Eastern Hill University, Shillong, Meghalaya, India

Abstract. The EHS3D international project in the Eastern Tibet and NE India brought in recent 15 years a huge array of magnetotelluric (MT) and magnetovariational (MV) soundings. These soundings were performed simultaneously in clusters containing several broadband (BMT) and long-period (LMT) field instruments and local geomagnetic observatories (Xiao et al. 2010; Varentsov et al. 2010). Reliably estimated impedances, tippers and horizontal MV responses made possible the resolution of conductivity structures within the whole tectonosphere at depths reaching 250 km. The model resulted from 2D+ inversion of joint MT/MV data set along the submeridional EHS-3 profile crossing the whole Eastern Tibet outlined sedimentary basins, subhorizontal upper crustal conductors, bright crustal-mantle conducting anomalies above dipping plates, well resistive lithospheric mantle and several "asthenospheric" cells. Recently, the EHS3D array has been extended with two new profiles. The EHS-4 profile fills the gap between EHS-2 and EHS-3 geotraverses in China, while the EHS-IND profile continues Tibetan profiles for more than 500 km into India through the Brahmaputra Valley. We present a view at the extended EHS3D data set and first 1D model along the joint profile from the Indian Craton to the SE Tibetan Region.

Keywords: Tibet · Himalaya · Simultaneous magnetotelluric soundings · Inversion · Geoelectric anomalies · Crustal flow

1 Introduction

The EHS3D international project was initiated by D. Bai and widely discussed in 2004 during the XVII EMIW in Hyderabad, India. First soundings were made in 2007 and 2009 by the Chinese team along two long (>1000 km) profiles (EHS-2 and -3, Fig. 1) in the Eastern Tibet in cooperation with Russian and Ukrainian scientists. 2D models based on BMT impedance data along EHS3D profiles were presented in (Bai et al.

© Springer Nature Switzerland AG 2019
A. A. Zhamaletdinov and Y. L. Rebetsky (Eds.): SPS 2018, SPEES, pp. 72–82, 2019.
https://doi.org/10.1007/978-3-030-35906-5_10

2010). These models outlined bright lateral variations in the crustal-mantle conductivity, but were strictly limited in the vertical resolution of the revealed anomalies. Several models along the EHS-3 profile based on the joint 2D+ inversion (Varentsov 2015a) of all long-period MT/MV data, described in (Varentsov et al. 2010; Varentsov and Bai 2015), were able to separate conducting structures at upper crustal, low crustal and "asthenospheric" levels. All the constructed models justified the need to extend observations at profile edges at the Chinese side and to add soundings over the border in NE India.

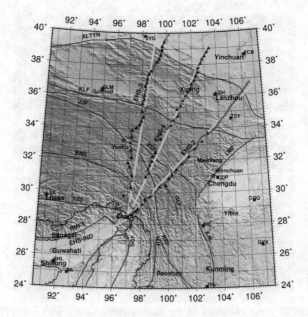

Fig. 1. The EHS3D simultaneous MT/MV sounding array in the Eastern Tibet: triangles – LMT sites, smaller circles – BMT sites, squares – geomagnetic observatories, red stars – recent strong earthquakes; tectonic lines – see (Bai et al. 2010).

During the second EHS3D project phase since 2010 more soundings were made at EHS-2,3 profiles, and a new EHS-4 profile was worked out between them (Fig. 1), with both BMT and LMT data collected as earlier.

Finally, in 2016 at the third EHS3D phase, the Indian-Russian subproject was started to make soundings along the EHS-IND profile in the Brahmaputra Valley. The profile was further extended to the total number of 30 sites in 2017–18 (Fig. 2, red triangles). In addition, results from several tens of previous Phoenix BMT soundings (Fig. 2, white diamonds) were integrated into the joint database. This paper presents first results in the analysis of integrated data sets from all three project phases.

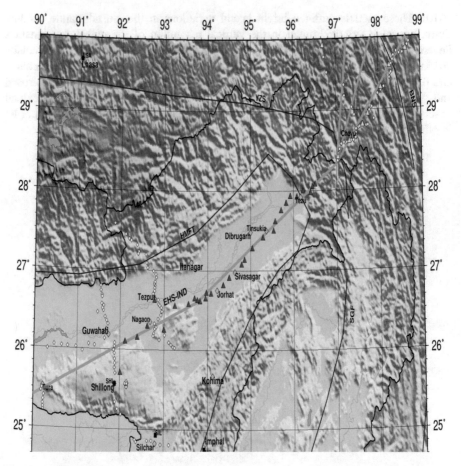

Fig. 2. The EHS3D array extension in the Brahmaputra Valley (NE India): triangles – new simultaneous LMT sites along the EHS-IND profile, smaller diamonds – BMT sites from previous studies, squares – geomagnetic observatories; tectonic lines – as in Fig. 1

2   MT/MV Data Acquisition and Processing

In China, LMT sites were collected with the LEMI-417, and the denser grid of BMT sites - with the Phoenix MTU instruments. The maximal joint period range at BMT +LMT sites achieves 0.003–12000 s. In India, the KMS-820 instruments were used to get data in the joint BMT/LMT period range of 0.125–8192 s, giving the penetration depth greater than 200 km.

The Russian multi-site data processing system (Varentsov 2015b) was applied to estimate impedance, tipper and horizontal MV responses at all LMT sites (including recent Indian KMS-820 observations) and at a part of BMT sites using 1 s data records. These estimations were supported with remote 1 s MV observations at a number of surrounding Chinese, Indian and Vietnamese geomagnetic observatories. The resistive subsurface structure in the most parts of the study region explains quite distant influence of industrial EM noises and justifies the need to apply multi-site data processing tools. At least 3–4 simultaneous remote sites were usually used in the multi-RR estimation. The horizontal MV responses were primarily estimated relative to the LZH observatory.

The total number of processed LMT sites exceeds 80 at the Chinese and 30 at the Indian sides. All Chinese and the most of Indian BMT sites were processed with the Phoenix SSMT2000 software. The impedance BMT and LMT data at common sites were compared and integrated in the period range of 0.1–10000 s. The MV datasets were limited to the range from 16–32 until 10000 s.

3 New TF Data Analysis

We mainly focus in this paper on the part of the EHS3D sounding array located south from the 30°E latitude (Fig. 2). The subsurface (sedimentary) conductance (Fig. 3), estimated asymptotically from the effective (square determinant) impedance at 16 s, reaches 400 S in the Upper Brahmaputra Basin (UBB) and is generally less than 100 S westwards. Figure 4 shows maps of extreme apparent resistivities (from the Spitz amplitude decomposition), while Fig. 5 gives maps of extreme phase tensor phases (from the CBB decomposition) at the 2048 s period. These maps definitely indicate the highly resistive deep crustal structure below the UBB continuing NE almost until the Chinese border in the contrast with sufficiently more conducting adjacent Tibetan blocks, and the medium conducting zone (Gokarn et al. 2008) around the Shillong Plateau (SHP). The highest resistivity is got for the crustal block located SW from the UBB. The SW edge of the UBB seems to be the most contrast, especially when looking at pseudo sections of the extreme impedance data (Fig. 6). The impedance principle directions are outlined in Fig. 7 with the phase tensor CBB extreme ellipses (90°-rotated) at the 512 s period.

Figure 8 presents complex diagrams of the invariant MV responses at the 2048 s period for NE India and the whole EHS3D area. Here the horizontal MV data are recalculated relative to the Shillong (SHL) geomagnetic observatory shown by the white star. These data are looking quite normal in NE India with minor inhomogeneities related to the currents within the UBB. The strongest anomaly in the horizontal response maximal amplitude (over 1.7, locally exceeding 1.9) appears from the BNS-NJF until JJF-XSF-XJF Tibetan fault zones. The anomalous maximal amplitudes extend further north at the level above 1.5 until the Altyn fault zone with a number of higher local anomalies. The strike of the highest amplitude zone reveals in long and narrow green ellipses, changing the maximal axis directions from SE to SSE on the way from the Eastern Tibet towards the Yunnan province. This zone follows the widely discussed crustal flow band, which, most probably, continues according to recent MT/MV studies (Nikiforov et al. 2018) until the Northern Vietnam territory and the Bacbo Bay (westwards from the Red River Fault). Real induction vectors (in Wiese notation) SW from this zone (Fig. 8, black arrows) naturally orient away from it. In NE India, these arrows are small enough; the largest of them at the SW edge of the UBB show western orientation, pointing away from deep conducting structures assumed somewhere eastwards close to the Myanmar border (Fig. 8).

76 Iv. M. Varentsov et al.

4 1D Inversion Models Along Joint EHS-IND and EHS-2 Profiles

Figure 9 finally displays a sequence of 1D resistivity models derived from the effective impedance (with the phase priority over the apparent resistivity) along the EHS-IND profile and the SW part of the EHS-2 profile. These images quantify in the first approximation sedimentary and crustal-mantle resistivity changes outlined in maps (Figs. 3, 4, 5 and 6) and pseudo sections (Fig. 7).

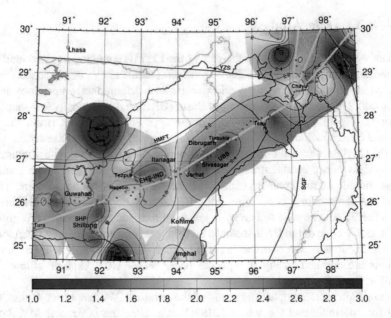

Fig. 3. The map of the subsurface (sedimentary) conductance (S, lg-scale) around the EHS-IND profile (estimated from the effective - square determinant impedance at 16 s period); UBB – the Upper Brahmaputra Basin, SHP – the Shillong Plateau, tectonic lines – as in Fig. 1

The whole section at sites DUP and JR2 is resistive above hundreds of $\Omega \cdot$ m. Sites within the UBB (JR4, DB2, TS2 and SD2) display conductivity anomalies in the sedimentary cover and the resistive structure below. First Chinese sites within the Lhasa block (202 and 208) outline moderate upper crustal conductors in the background of resistive subsurface and deeper crustal-mantle layers. The next site 220 indicates the general conductivity increase below the 20 km depth. Finally, the last 228 site, located north from the BNS fault zone and representing the "crustal flow" belt, reveals in this depth range resistivities well below 10 $\Omega \cdot$ m.

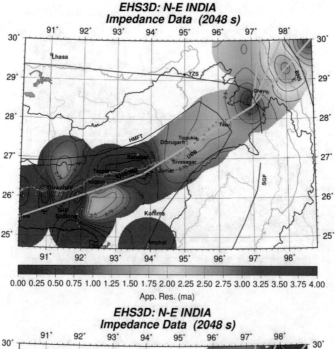

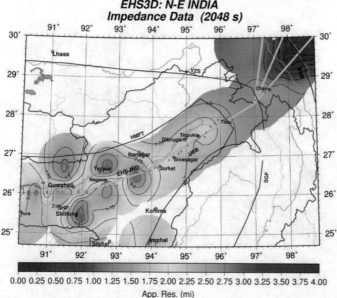

Fig. 4. The impedance invariants around the EHS-IND profile: maximal (top) and minimal (bottom) apparent resistivities (Spitz amplitude decomposition, $\Omega \cdot$ m, lg-scale, 2048 s); legend – as in Fig. 3

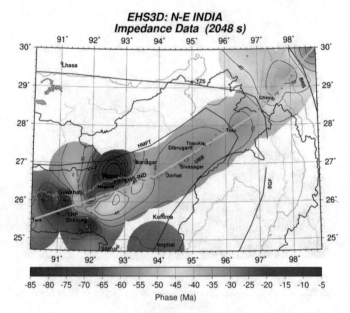

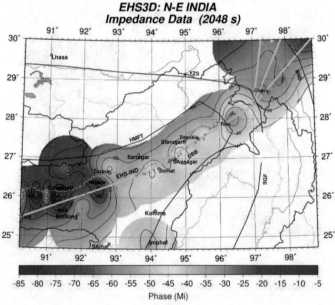

Fig. 5. The impedance invariants around the EHS-IND profile: maximal (closest to 0°, top) and minimal (closest to −90°, bottom) CBB phases (deg., convention with [−90°, 0°] 1D quadrant, 2048 s)

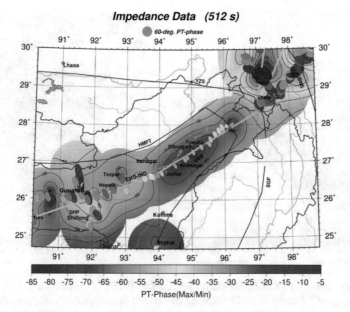

Fig. 6. Diagrams of TF invariants around the EHS-IND profile: CBB extremal phase ellipses over the maximal phase map (90°-rotated, filled with minimal phase color, conventions as in Fig. 5, 512 s)

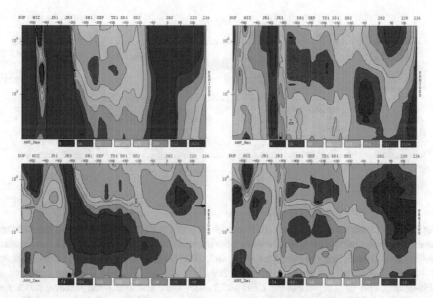

Fig. 7. Pseudo sections of the extremal apparent resistivities (top panels, Spitz decomposition, $\Omega \cdot m$, lg-scale) and CBB phases (bottom panels, deg., convention as in Fig. 5) along the EHS-IND and EHS-2 continued profiles: left column – maximal invariants, right column – minimal invariants; horizontal axis – profile coordinates (positive in China and negative in India); vertical axis – periods (s, lg-scale)

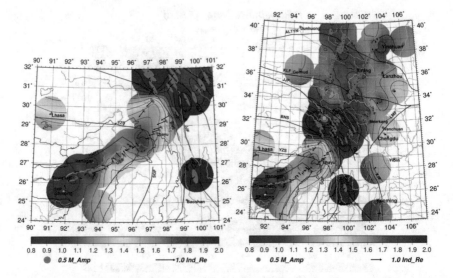

Fig. 8. Diagrams of MV data invariants around the EHS-IND profile: left – the maximal amplitude map of the horizontal MV response (relative to the SHL observatory) overlapped with green extremal amplitude ellipses for its anomalous part (90°-rotated, 2048 s) and black real induction arrows (Wiese convention, same period); right – the same diagram for the whole EHS3D region; legend – as in Fig. 3

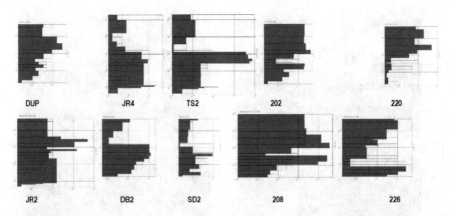

Fig. 9. The comparison of 1D resistivity models at profiles EHS-IND and EHS-2 derived from the effective impedance data with the phase priority; horizontal axis – resistivity, $\Omega \cdot m$, lg-scale; vertical axis – depth, km, lg-scale; sites DUP, JR2, JR4, DB2, TS2, SD2 are successively located at the EHS-IND from SW to NE; sites 202, 208, 220, 226 follow them from the border at the EHS-2; sites JR4-SD2 represent the UBB block, sites DUP and JR2 – the resistive block next from SW, sites 202–220 represent the Lhasa block, the site 226 enters the strongest horizontal MV amplitude anomaly north from BNS (Fig. 8)

5   Conclusions

New BMT+LMT data obtained at the Chinese profiles fill gaps in the Tibetan part of the EHS3D array, while new Indian soundings extend it far to SW. The data collected in NE India look quite consistent along the EHS-IND profile. However, they have medium to strong 3D distortions in almost all components, and their principle directions change with periods. Their invariant analysis is to be continued in more details to properly select datasets for the joint 2D+ MT/MV inversion along integrated geotraverses, being longer than 2000 km and completely covering the Tibetan section with adjacent platform edges. At the same time, the achieved EHS3D array extension gives good grounds to start 3D data interpretation.

The existence of resistive crustal-mantle blocks revealed around the Upper Brahmaputra Basin explains the general geodynamic stability of the EHS (Eastern Himalaya Syntaxes) zone, while changes in this block structure outline areas of possible increased seismicity. The strongest horizontal MV anomalies most reliably trace continuous crustal-mantle conductors in the SE Tibetan part and image the location of the proposed crustal flow from the Tibetan Plateau into the Indochina Region.

Acknowledgments. The Russian RFBR grants 16-55-45052_IND, 17-55-53102_GFEN and the Indian DST grant INT/RUS/RFBR P-247 partly supported this research. The authors are grateful to all members of the EHS3D Working Group. We also thank the teams from the INTER-MAGNET, CEA and IIG networks for very useful 1 s MV time series from a number of surrounding geomagnetic observatories.

References

Bai, D., Unsworth, M., Meju, M., et al.: Crustal deformation of the eastern Tibetan Plateau revealed by MT imaging. Nat. Geosci. (2010). https://doi.org/10.1038/ngeo830

Gokarn, S.G., Gupta, G., Walia, D., et al.: Deep geoelectric structure over the Lower Brahmaputra Valley and Shillong Plateau, NE India using magnetotellurics. Geophys. J. Int. **173**, 92–104 (2008). https://doi.org/10.1111/j.1365-246X.2007.03711.x

Nikiforov, V.M., Shkabarnya, G.N., Kaplun, V.B., et al.: Electroconducting elements of the ultradeep fluid–fault systems as indicators of seismically active zones of the eastern margin of the Eurasian Continent: evidence from MT data. Dokl. Earth Sci. **480**(2), 832–839 (2018). https://doi.org/10.1134/S1028334X1806028

Varentsov, Iv.M.: Methods of joint robust inversion in MT and MV studies with application to synthetic datasets. In: Spichak, V.V. (ed.) Electromagnetic Sounding of the Earth's Interior Theory, Modeling, Practice, pp. 191–229. Elsevier (2015a)

Varentsov, Iv.M.: Arrays of simultaneous EM soundings: design, data processing, analysis, and inversion. In: Spichak, V.V. (ed.) EM Sounding of the Earth's Interior: Theory, Modeling, Practice, pp. 271–299. Elsevier (2015b)

Varentsov, Iv.M., Bai, D.: Geoelectrical model of the tectonosphere of the Eastern Tibet from the data of deep and prospecting MT/MV soundings. In: Problems of Geodynamics and Geoecology (Proceedings VI International Symposium), pp. 169–177. NS RAS, Bishkek (2015). (in Russian)

Varentsov, Iv.M., Bai, D., Sokolova, E.Yu.: Joint inversion of long-period MT/MV data at EHS3D transects (Eastern Tibet). In: XX EM Induction Workshop (Ext. Abstracts). Egypt, Giza, p. S7-05 (2010)

Xiao, P., Bai, D., Lui, M., et al.: Study on long-period MT sounding - the LMT transfer function in eastern Tibetan Plateau. Seismology and Geology. **32**(1), 38–50 (2010). https://doi.org/10.3969/j.issn.0253-4967.2010.01.004. (In Chinese)

The Influence of Anomalous Magnetic Permeability on MT/MV Responses Observed at the Voronezh Massif Near Intensive Magnetic Anomalies

I. N. Lozovsky[1]([✉]), Iv. M. Varentsov[1], and R.-U. Börner[2]

[1] Geoelectromagnetic Research Center,
Schmidt Institute of Physics of the Earth RAS, Moscow, Troitsk, Russia
i.n.lozovsky@yandex.ru, ivan_varentsov@mail.ru
[2] Institute of Geophysics, TU Bergakademie Freiberg, Freiberg, Germany

Abstract. The basic medium model considered in magnetotelluric (MT) and magnetovariational (MV) studies assumes the magnetic permeability being fixed everywhere at the free space level. However, this assumption can cause distortions in the MT sounding results near strong permanent magnetic field anomalies. We made 2D modelling of such distortions arising along the KIROVOGRAD project MT/MV sounding profiles at the western slope of the Voronezh Massif. Minor changes are seen in modelled data for TM mode components. At the same time, the MV anomalies in TE mode are decreasing with the increase of magnetic permeability within upper crustal magnetic blocks. This effect may cause the decrease (or even break) of deeper low crustal conductive anomalies just below strongly magnetic bodies.

Keywords: Magnetotellurics · Geoelectric anomalies · Magnetic
permeability · Voronezh Massif

Modelling and inversion methods applied in magnetotellurics with very few exceptions, like (Dobrokhotova and Yudin 1981; Mukherjee and Everett 2011), account for the medium model with the vacuum magnetic permeability μ_0. However, for highly magnetized rocks the magnetic permeability value can be several times greater. Thus, MT/MV responses collected in the vicinity of intensive magnetic anomalies can be distorted due to the unaccounted effect of anomalous magnetic permeability leading to false interpretation of obtained geoelectric models (Dobrokhotova and Yudin 1981).

The KIROVOGRAD project MT/MV array was worked out in 2007–16 to study the geoelectric structure of the lithosphere at the western slope of the Voronezh Massif (Varentsov et al. 2012; Aleksanova et al. 2013; Varentsov 2015; Kulikov et al. 2018). More than 220 simultaneous broadband and long-period MT/MV soundings have covered a wide territory of 49–55°N, 31–37°E (Fig. 1).

© Springer Nature Switzerland AG 2019
A. A. Zhamaletdinov and Y. L. Rebetsky (Eds.): SPS 2018, SPEES, pp. 83–88, 2019.
https://doi.org/10.1007/978-3-030-35906-5_11

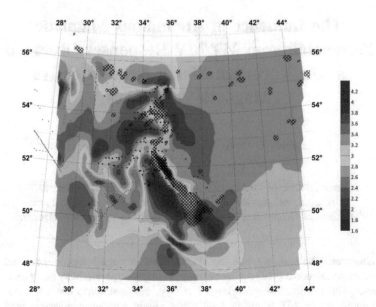

Fig. 1. The crustal conductance map (S, lg-scale) from quasi-3D inversion of horizontal MV responses with black contours of strong permanent magnetic anomalies (>800 nT); green dots show sites of simultaneous MT/MV soundings.

The study area is characterized by the high level of industrial noise, thus multisite robust schemes (Varentsov 2015) were applied for the reliable estimation of MT/MV transfer functions, namely, impedance, tipper and horizontal magnetic response. 1D and 2D+ inversion procedures were applied at a number of profiles (Varentsov et al. 2012; Aleksanova et al. 2013) and the whole array data were inverted using quasi-3D (Varentsov et al. 2017) and full 3D (Kulikov et al. 2018) inversion techniques. The inversion models have outlined Kirovograd, Kursk and Kirov-Baryatino conductive anomalies in the lower crust and less conductive subvertical zones above them in the upper crust.

The research region is characterized with the presence of very intensive Kursk and Baryatino magnetic anomalies (Abramova et al. 2012; Aleksanova et al. 2013). The nature of these bright anomalies relates to ferrous quartzite formations with anomalously high values of magnetic permeability located in the upper crystalline basement. Figure 1 shows that contours of strong magnetic anomalies correlate well with the contours of crustal conductive structures (Varentsov et al. 2017).

To study the influence of anomalous magnetic permeability on the MT/MV data the forward 2D modelling code based on the finite element algorithm for adaptive unstructured triangular grids (Franke et al. 2007) was applied. This code was successfully used in the similar modelling study in (Szarka et al. 2010). The model prepared for our study (Fig. 2c) reflects principle features of 2D+ inversion geoelectric structure (Fig. 2b) obtained at the ZHIZDRA profile (Varentsov et al. 2012) and the observed magnetic anomalies (Fig. 2a).

The model includes the subhorizontal crustal conductor (with $10 \ \Omega \cdot m$ resistivity, 100 km width, 20 km upper edge depth and 10 km thickness), two less conducting

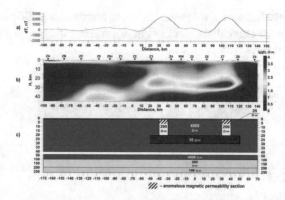

Fig. 2. The ZHIZDRA profile: a – permanent magnetic field (EMAG2 data, nT); b – resistivity section ($\Omega \cdot$ m, lg-scale) from 8-component 2D+ joint MT/MV data inversion; c – resistivity model prepared for 2D forward modelling study with anomalous magnetic permeability sections highlighted.

subvertical zones above this structure (with 200 $\Omega \cdot$ m resistivity, 10 km width, 1 km upper edge depth and 19 km thickness) and the layered background section with crystal resistivity of 4000 $\Omega \cdot$ m. Blocks with anomalous magnetic permeability were located only in the upper part of subvertical conducting zones (with 1 km upper edge depth, 5 km thickness) as outlined in Fig. 2c. The magnetic permeability values for these blocks were selected from the sequence of μ_0, $1.5 \cdot \mu_0$, $3 \cdot \mu_0$ or $5 \cdot \mu_0$ alternatives. The modelling was held at sufficiently dense spatial grid with the use of Dirichlet boundary conditions for periods selected in 2^n sequence in the range of 0.125–8192 s.

The modelled electric fields, namely, E_x, in TM mode and E_y in TE mode, are insignificantly influenced by the magnetic permeability anomalies (Fig. 3).

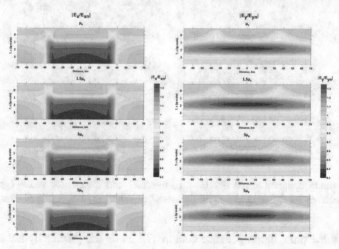

Fig. 3. Pseudo sections of electric field components for different magnetic permeability values (1, 1.5, 3 and 5 $\cdot$ times μ_0); profile coordinates (in km) - at the horizontal axis, periods (in s, lg-scale) - at the vertical axis.

On contrary, the horizontal magnetic (H_x) field amplitudes in TE mode significantly decrease within such anomalous blocks (Fig. 4), forcing the overestimation of correspondent apparent resistivity (Fig. 5).

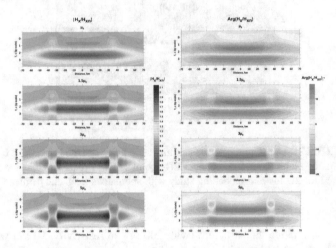

Fig. 4. TE mode horizontal MV response pseudo sections (amplitude and phase) calculated for different magnetic permeability values (1, 1.5, 3 and 5 · times μ_0).

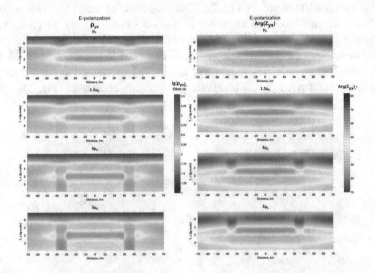

Fig. 5. TE mode apparent resistivity ($\Omega \cdot$ m, lg-scale) and impedance phase (grad.) pseudo sections calculated for different magnetic permeability values (1, 1.5, 3 and 5 · times μ_0).

At the same time, the impedance data in TM mode are practically not sensitive to magnetic permeability variations (Fig. 6).

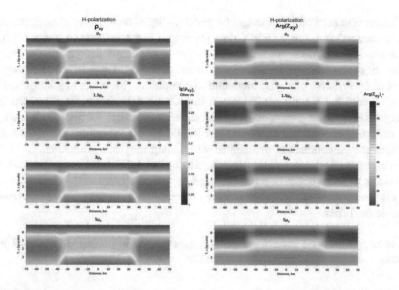

Fig. 6. TM mode apparent resistivity ($\Omega \cdot$ m, lg-scale) and impedance phase (grad.) pseudo sections calculated for different magnetic permeability values (1, 1.5, 3 and 5 $\cdot$ times μ_0).

In the vertical magnetic field component (H_z) and in the tipper data the magnetic permeability anomalies cause false structures of different signs (Fig. 7). All the MV data distortions grow with the increase of the magnetic permeability.

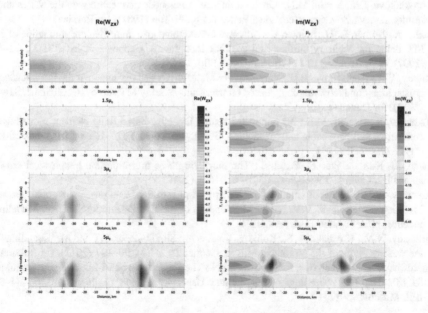

Fig. 7. TE mode tipper pseudo sections (Re and Im) calculated for different magnetic permeability values (1, 1.5, 3 and 5 $\cdot$ times μ_0).

The unaccounted effect of magnetic permeability variations can distort geoelectric models obtained from the MT/MV data inversion. The conductance under the blocks with anomalous magnetic permeability could be significantly decreased. The image of continuous subhorizontal conductive structures below bright magnetic bodies may even be broken by resistive distorted fragments. To minimize this effect in joint bimodal 2D data inversion, the weight of TM mode impedance data should be increased in the vicinity of permanent magnetic field anomalies, especially, for the phase data. The same distorting effect could also take place in 3D inversion solutions along the strike of dominant crustal conductors overlapped with strong magnetic anomalies.

To understand and partly eliminate the effect of anomalous magnetic permeability it is advisable to hold similar or more extended modelling studies in the areas with strong magnetic anomalies and to develop joint inversion tools for MT/MV and permanent magnetic field data.

Acknowledgments. This research was partly supported by the RFBR grant 16 35-00529_mol_a. The authors are grateful to all members of the KIROVOGRAD WG.

References

Abramova, D.Yu., Abramova, L.M., Varentsov, Iv.M., et al.: Correlation of permanent magnetic field anomalies and crustal geoelectric structures at the western slope of the Voronezh Massif. Geofiz. Zh. **34**(4), 62–69 (2012). (in Russian)

Aleksanova, E.D., Varentsov, Iv.M., Kulikov, V.A., et al.: Deep conductivity anomalies in the northern part of Voronezh Anticlise. Geofizika **2**, 31–37 (2013). (in Russian)

Dobrokhotova, I.A., Yudin, M.N.: On the influence of magnetic permeability on the MT method results. Izv. VUZov. Ser. Geologia i Razvedka **6**, 99–106 (1981). (in Russian)

Franke, A., Börner, R.-U., Spitzer, K.: Adaptive unstructured grid finite element simulation of 2D MT fields for arbitrary surface and seafloor topography. Geophys. J. Int. **171**(1), 71–86 (2007). https://doi.org/10.1111/j.1365-246X.2007.03481.x

Kulikov, V.A., Aleksanova, E.D., Varentsov, Iv.M., et al.: Resistivity image of the Baryatinskaya crustal high-conductive anomaly based on the results of array MT survey. Geofizika **1**, 31–43 (2018). (in Russian)

Mukherjee, S., Everett, M.E.: 3D controlled-source EM edge-based finite element modeling of conductive and permeable heterogeneities. Geophysics **76**(4), F215–F226 (2011). https://doi.org/10.1190/1.3571045

Szarka, L., Kiss, J., Prácser, E., et al.: The magnetic phase transition and geophysical crustal anomalies. Chinese J. Geophys. **53**(3), 612–621 (2010)

Varentsov, Iv.M.: Arrays of simultaneous EM soundings: design, data processing, analysis, and inversion. In: Spichak, V.V. (ed.) EM Sounding of the Earth's Interior: Theory, Modeling, Practice, pp. 271–299. Elsevier (2015)

Varentsov, Iv.M., Kovacikova, S., Kulikov, V.A., et al.: Simultaneous MT and MV soundings at the western slope of the Voronezh Massif. Geofiz. Zh. **34**(4), 90–107 (2012). (in Russian)

Varentsov, Iv.M., Kovacikova, S., Lozovsky, I.N.: The advances in quasi-3D inversion technique for MV data. In: Proceedings of International Uspensky Seminar (44th session), pp. 84–89. IPE, Moscow (2017)

CSAMT and MT-AMT Static Shift Correction by Means of the Total Horizontal Magnetic Field

A. A. Zhamaletdinov[1,2(✉)], A. A. Skorokhodov[2],
and A. N. Shevtsov[2]

[1] St. Petersburg Brunch of IZMIRAN, St. Petersburg, Russia
abd.zham@mail.ru
[2] Geological Institute KSC RAS, Apatity, Russia

Abstract. "Static shift" appear as the frequency-independent influence on results of EM soundings made with natural or controlled sources. Static shift distortion displaces apparent resistivity curves usually down for several orders and appear as the main reason of false results. In this paper we propose the quantitative method of "static shift" correction. The use of the total horizontal magnetic field measurements in the field of controlled source frequency sounding is the base of this technique. The use of magnetic measurements with induction coils, free of galvanic connection with the ground, made possible to calculate quantitatively the static shift distortions. Experiment on control source sounding with spacing up to 25 and 50 km between transmitter and receiver is implemented for to study the static shift in the Kovdor area (Kola Peninsula). Apparent resistivity curves, calculated by the total electric field (ρE_{tot}) happened to be shifted in some measuring points down up to 300–500% compare to apparent resistivity curves calculated by the total magnetic field (ρH_{tot}). Impedance curves (ρZ_{tot}) shifted dawn up to 900–2500% compare to ρH_{tot}. After corrections the anomalous lowering of apparent resistivity disappeared and the area of Yona-Belomorian craton happened to be homogeneous in accordance with geological presupposition.

Keywords: Static shift · Control source soundings · Magnetotelluric soundings · Kola Peninsula

"Static shift" distortions are the very old painful point of electromagnetic (EM) soundings. They appear due to influence from the near to surface. There are several ways to avoid "static shift" distortions, invented and applied by many researchers, mentioned below.

1. The use of MN measuring lines that have the lengths exceeding dimensions of the near to surface heterogeneities (Berdichevsky 1968). This technique is of the most effective application on the territories of crystalline shields, where static shifts mostly appear due to mosaic like moraine cover of rather thin thickness (Zhamaletdinov 1990).

© Springer Nature Switzerland AG 2019
A. A. Zhamaletdinov and Y. L. Rebetsky (Eds.): SPS 2018, SPEES, pp. 89–94, 2019.
https://doi.org/10.1007/978-3-030-35906-5_12

2. The parallel displacement of magnetotelluric sounding (MTS) apparent resistivity curves till coincidence of their long-period wings with apparent resistivity curves of magnetovariational global sounding data (MVS) (Rokityansky 1971).
3. Static shift correction by means of the OCCAM technique (Groot-Hedlin and Constable 1990) based on the use of phase curves as dominant for to correct the MTS apparent resistivity curves.
4. Statistical method based on the joint analysis of amplitude and phase curves of MTS on the sufficiently large array of observations (Jones 1988).
5. Application of induction soundings with magnetic loops (TDEM) for to correct high frequency parts of MTS curves (Feldman 1994).
6. Separation of magnetotelluric field on the local and regional parts (decomposition technique) (Groom and Bailey 1989).

Many other approaches exist for to avoid "static shift" distortions, for example phase tensor technique (Caldwell et al. 2012) and others. Without analysis of advantages and disadvantages of different methods aimed to combat the "static-shift", we can note here that all they are of qualitative character.

In this paper we propose the quantitative method of "static shift" correction. The use of the total horizontal magnetic field measurements in the field of controlled source frequency sounding is the base of this technique. Theoretical basement is illustrated below on example of the field of grounded dipole in a quasi-stationary (plane-wave) zone (Vanyan 1965). Dipole is oriented along the X axis. Equations are presented for intensity of longitudinal electric field E_x (1), transversal magnetic field H_y (2) and for the input impedance $|Z_{xy}|$ (3).

$$E_x = \frac{I \cdot L_{AB} \cdot (3 \cdot \cos^2 \theta - 2)}{2 \cdot \pi \cdot r^3} \cdot \rho, \ [V/m]; \tag{1}$$

$$H_y = \frac{I \cdot L_{AB} \cdot (3 \cdot \cos^2 \theta - 2)}{2 \cdot \pi \cdot r^3 \cdot \sqrt{\omega \mu_0}} \cdot \sqrt{\rho}, \ [A/m], \tag{2}$$

$$|Z_{xy}| = \left| \frac{E_x}{H_y} \right| = \sqrt{\omega \mu_0 \rho}, \ [Ohm]; \tag{3}$$

From these expressions it can be seen that electrical resistivity of the lower half space in the field of grounded electric dipole can be defined both from electric (1) and from magnetic (2) fields separately and from their relations, if to use the input impedance (3). The difference between relations (3) and (1–2) is that intensity of electric field E_x depends on the electrical resistivity ρ linearly whereas intensity of magnetic field H_y and the input impedance Z_{xy} depends on electrical resistivity ρ under the square root. Another characteristic property of Eqs. (1–3) is, that the input impedance Z_{xy} depends on $\sqrt{\rho}$ and $\sqrt{\omega}$ the only, and don't depends on other parameters of the source – intensity of current and geometry of the source. That property is true only in the quasi-stationary (plane wave) conditions and is equitable for the any type of source. This consequence is commonly used in magnetotellurics in frame of "Tikhonov-Kagniard" model. Magnetic H_y and electric E_x fields can be used for the apparent resistivity calculation only in the field of controlled sources, as they depend on geo-metric parameters and intensity of current. Another feature is that the magnetic sensor is insulated from the ground. By the reason it is free of influence from static shift influence and can by used for the qualitative estimation of static shift distortions, affected on electric and impedance parameters.

Experimental verification of this technique of static shift effect estimation has been implemented while performing the multipath control source frequency sounding syn-chronously with AMT-MT sounding (experiment "Kovdor-2015").The soundings in the experiment "Kovdor-2015" were performed in frame of homogeneous high resis-tive Yona-Belomorian craton of Archaean age (Fig. 1).

Two transmitting systems were situated at distance of 85 km between each other (Fig. 1). Each transmitting system consists of two mutually quasi-orthogonal grounded electric dipoles of 1.5–2.0 km length in average (L1–L2) and (L3–L4). Car-generator of 29 kW power has been used as the source of electromagnetic field in the frequency range of 4 Hz–2 kHz (Kolobov et al. 2018). Receivers in the "Kovdor-2015" experi-ment (11 sites) were situated at distances of 25 and 50 km from transmitters. The measurements at each site were carried out with the use two stations operating simultaneously - of 24 bit digital station KVVN-7 and 32 bit digital station VMTU-10. The application of two measuring stations at each receiving site was implemented for to improve the reliability of results and to increase stability of the inverse problem solution.

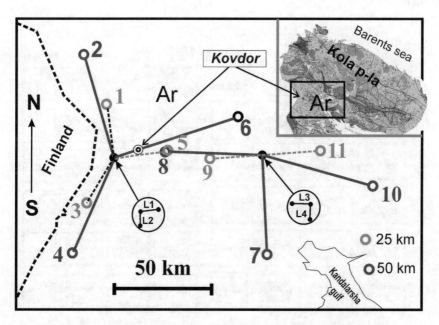

Fig. 1. Transmitting and receiving lines of the experiment "Kovdor-2015"

Results of soundings, obtained with the use of two different measuring stations and with the use of two quasi-orthogonal electrical dipoles, are in a good coincidence between each other. The averaged scattering is in the scope of 10–15% (Fig. 2). Apparent resistivity has been calculated by the modulus of the total horizontal magnetic and electric fields. Below Eq. (4) are presented for calculation of the apparent resistivity in the field of horizontal grounded electrical dipole by the total horizontal magnetic field

$$\rho_{tot}^{H} = 2\pi f \cdot \mu_0 \cdot \left(K_{tot}^{H} \cdot \frac{H_{tot}}{I \cdot L_{AB}} \right)^2, \dots H_{tot} = \sqrt{H_x^2 + H_y^2}$$

$$K_{tot}^{H} = \frac{K_x^{H} \cdot K_y^{H}}{\sqrt{\left(K_x^{H}\right)^2 + \left(K_y^{H}\right)^2}}, \dots K_x^{H} = \left(\frac{2\pi \cdot r^3}{3 \cdot Cos\,\theta \cdot Sin\,\theta} \right)^2,$$

$$K_y^{H} = \left(\frac{2\pi \cdot r^3}{3Cos^2\theta - 2} \right)^2 \tag{4}$$

In frame of quasi-stationary zone (200 Hz–0.5 kHz) the apparent resistivity curves calculated from the total magnetic field ρ_{tot}^{H} happened to be free of influence from the "static-shift" (Fig. 2). Significances of apparent resistivity ρ_{tot}^{H} are around of about 10^4 Ω m.

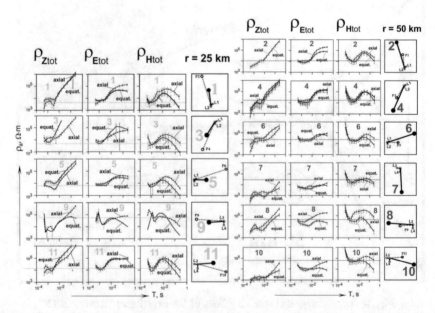

Fig. 2. The common diagram of apparent resistivity curves from the "Kovdor-2015" experiment. Left panel shows results of CSAMT soundings (ρZ_{tot}, ρE_{tot}, ρH_{tot}) at distances of about 25 km. The rite panel is same, but for distances of about 50 km

Static shift" distortions have been noticed at four sites (7, 8, 9, 10) by electric and impedance apparent resistivity cures (Fig. 2). The use of magnetic measurements with induction coils, free of galvanic connection with the ground, made possible to calculate quantitatively the static shift distortions. Apparent resistivity curves, calculated by the total electric field (ρE_{tot}), are shifted down up to 300–500% compare to apparent resistivity curves calculated by the total magnetic field (ρH_{tot}). Impedance curves (ρZ_{tot}), in accordance with Eqs. 1–3, are shifted dawn up to 900–2500% compare to ρH_{tot}. After these corrections the anomalous lowering of apparent resistivity up to 10–20 times (from 10–20 kΩ m till 10^3 Ω m) disappeared and the area of Yona-Belomorian craton happened to be homogeneous in accordance with geological presupposition.

CSAMT sounding results are presented separately for the site 7, where static shift distortion is especially great (Fig. 3a). This result is compared with MT sounding data obtained at the same site at the next year (Fig. 3b).

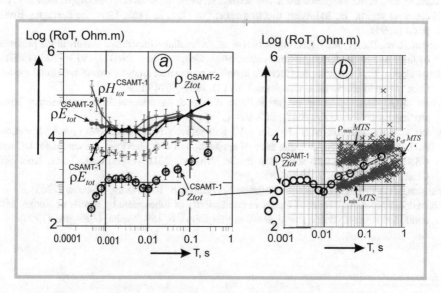

Fig. 3. Comparison of CSAMT and MT sounding results at the site 7. Explanations to the Fig. 3 given in the text

MTS apparent effective resistivity curve ($\rho_{eff}MTS$) on the Fig. 3b coincides with the measured CSAMT apparent resistivity curve ($\rho_{Ztot}^{CSAMT-1}$), calculated by the total impedance. Comparison with the measured total magnetic apparent resistivity curve ($\rho_{Htot}^{CSAMT-1}$) shows that the curve $\rho_{Ztot}^{CSAMT-1}$ is shifted dawn on about 1.2 order. Corrected impedance curve $\rho_{Etot}^{CSAMT-2}$ is shown on the Fig. 3a. Apparent resistivity curve measured by the total electric field ($\rho_{Ztot}^{CSAMT-1}$) is shifted dawn on 0.5 order. Corrected curve ($\rho_{Etot}^{CSAMT-2}$) coincides with other curves in conditions of quasi stationary wave zone (200–700 Hz.). Presented results show, that MTS data ($\rho_{min}MTS$, $\rho_{max}MTS$ and $\rho_{eff}MTS$) should be shifted on a 1.2 order up for to correct static shift distortion.

Acknowledgments. Authors are grateful for the participation in the field work and the processing of materials to their colleagues – V. E. Kolesnikov. T. G. Korotkova, V. V. Kolobov, V. V. Ivonin, M. A. Birulya. This work was supported by the grant RFBR-18-05-00528.

References

Рокитянский И.И. Глубинные магнитотеллурические зондирования при наличии искажений от горизонтальных неоднородностей. Геофизический сборник. Киев, Наукова думка, вып. 43, 71–78 (1971)

Berdichevsky, M.N.: Elektricheskaya Razvedka Metodom Magnitotelluricheskogo Profilirovaniya (Electric Survey Using the method of Magnetotelluric Profiling). Nedra, Moscow (1968)

Caldwell, T.G., Bibby, H.M., Brown, C.: The magnetotelluric phase tensor. Geophys. J. Int. **191** (3), 1129–1134 (2012)

Feldman, I.S., et al.: Magnetotelluric and seismic study of the earth crust and upper mantle in the caucasus region. In: Induction Electromagnetique Dans la Terre. Univ. de Bretagne, Brest, p. 65 (1994)

Groom, R.W., Bailey, R.C.: Decomposition of magnetotelluric impedance tensors in the presence of local tree-dimensional galvanic distortion. J. Geophys. Res. **94**(B2), 1913–1925 (1989)

Groot-Hedlin, C., Constable, S.: Occam's inversion to generate smooth, two-dimensional models from magnetotelluric data. Geophysics 55(12), 1613–1624 (1990)

Jones, A.G.: Static shift of magnetotelluric data and its removal in a sedimentary basin environment. Geophysics 53(7), 967–978 (1988)

Kolobov, V.V., Barannik, M.B., Ivonin, V.V., Selivanov, V.N., Zhamaletdinov, A.A., Shevtsov, A.N., Skorokhodov A.A.: Experience of application of the "Energy-4" generator for DC and CSAMT electromagnetic soundings in the "Murman-2018" experiment. Trans. Kola Sci. Center RAS **9**(8(17)), 7–20 (2018)

Vanyan, L.L.: Principles of Electromagnetic Soundings, p. 107. Nedra, Moscow (1965)

Zhamaletdinov, A.A.: Model of electrical conductivity of lithosphere by results of studies with controlled sources (Baltic shield, Russian plateform), p. 159. Nauka, Leningrad (1990)

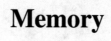

Memory

Professor Aida Kovtun – Great Scientist and Teacher in Geoelectrics

A. A. Zhamaletdinov[1(✉)], S. V. Buldyrev[2], I. L. Vardanyants[3],
N. I. Uspensky[3], and E. Yu. Sokolova[4]

[1] St. Petersburg Brunch of IZMIRAN, St. Petersburg, Russia
abd.zham@mail.ru
[2] Yeshiva University, New York, USA
buldyrev@yu.edu
[3] St. Petersburg State University, St. Petersburg, Russia
[4] Schmidt Institute of Physics of the Earth of the RAS, Moscow, Russia

In her youth Aida Andreyevna Kovtun wanted to be a historian, her favorite hero was Grand Duke Alexander Nevsky. But her father, Andrei G. Kovtun, a Communist, a teacher at the Military Academy of Logistics and Transport, a surveyor and cartographer, the kindest and modest man, advised her against this occupation. Aida - he named her so, because he was a big opera lover, - had significant talent for mathematics and natural sciences. Thus, she became a physicist. But Aida Kovtun interest in history lasted her whole life. Already in her mature years, she saw an extraordinary dream, as if somewhere in an infinite, light-filled space, an endless chain of women was weaving a glittering white canvas that stretched beyond the horizon. And there was an enormous joy in that. Waking up, she realized that the canvas is life, and its endlessness is the continuity of generations, and the meaning of human life is to weave this cloth so that it will not be interrupted. Therefore, she kept diaries and wrote memoirs about the family, about life, about science. At the same time, her main occupation was always her science - geoelectrics and her favorite department, the interests of which she lived.

Aida Kovtun graduated from the Physics Faculty of the Leningrad State University in 1953. Since 1957, Aida Andreyevna has gone a long way of scientific and peda-gogical activity at the Department of Physics of the Earth in the Saint Petersburg (former Leningrad) State University. Her works on the theory and technique of the data interpretation in magnetotellurics have received wide recognition in Russia and abroad.

A. A. Zhamaletdinov and Y. L. Rebetsky (Eds.): SPS 2018, SPEES, pp. 97–102, 2019.
https://doi.org/10.1007/978-3-030-35906-5_13

She investigated the properties of the asthenospheric layer under the Russian platform, studied the Ladoga anomaly of electrical conductivity, and traced the intermediate conducting layer in the middle part of the Earth's crust throughout the Baltic Shield. Aida Kovtun always distinguished herself by a variety of scientific interests, benevolence and inexhaustible devotion to her beloved research topic - magnetotellurics. From the very beginning worked with such Leningrad coryphaei as Boris Mikhailovich Yanovsky, Georgy Vasiljevich Molochnov, Boris Evgenievich Brunelli and others. In turn Aida Kovtun was always surrounded by talented students and followers, who were endlessly devoted, like she was, to her favorite research area - studying the electrical conductivity of the Earth's crust and the Upper Mantle using electromagnetic methods. Figure 1 shows a part of her team at the Baku Electromagnetic School in 1981. Aida Andreyevna herself is not in the photo, probably at this time she was busy preparing her lecture. But you can see her associates and students: Marina Dobrovolskaya, Lyudmila Porokhova, Marina Sholpo, Isabella Vardanyants, Slava Vagin. In the center is the Head of the Physics Faculty Professor Georgy Molochnov.

Fig. 1. The team of Aida Kovtun among participants of All the Soviet Union Electromagnetic Workshop in Baku, Azerbaidjan, October, 1981. From the right to the left: S. Vagin, V. Shuman, I. Vardanyants, professor G. Molochnov, M. Sholpo, L. Porokhova, M. Dobrovol'skaya, N. Chicherina, L. Galichanina, S. Kulik, Diana – doughter of A. Zhamaletdinov, who is taking this photo

The main distinguishing feature of Aida Andreyevna's creative activity was her ability to combine theory and practice. She studied with equal interest the deep electrical conductivity of the Russian Platform and the Baltic Shield and the development of numerical methods for modeling horizontally inhomogeneous media and the natural magnetic field of the Earth and the geothermal regime of the lithosphere. With her direct participation in 1959 in Borok Observatory, magnetotelluric curve of apparent resistivity, one of the first in the world, has been obtained. Under her leadership, a large volume of magnetotelluric research was carried out in the North-West of the East European platform.

It should be noted that the first stages of Aida Andreyevna's activity (50th and 60th of the last century) belong to the beginning of the magnetotelluric research in geophysics. Only low-frequency quartz magnetometers of the Bobrov's system (1 Hz and below) and magnetic variometers of Brunelli's system (10 Hz and below) were available to the researchers at that time. The main tools were photo paper and a logarithmic ruler. Interpretation of results at first was also limited to formal approaches based on the one-dimensional Tikhonov-Cagniard's model. Colleagues recall how Aida Andreyevna could spend a night with the rolls of photo paper and in the morning present a new apparent resistivity curve and a new deep electrical section of the Earth.

The first theoretical works of Kovtun appeared in the early 60s. They are devoted to the study of the behavior of the magnetotelluric field above the conducting wedge, the study of the ring current field on the surface of a multi-layered Earth, the processing and interpretation of MTS data, including the data accounting for horizontal inhomogeneity of the low half space. In the 70s, the main object of her interest in the field of experiment was the study of the structure of the Russian platform - a sedimentary cover and upper mantle. Results of all these studies are reflected in the textbook "Using the natural electromagnetic field in the study of the electrical conductivity of the Earth" (Kovtun 1980), which became widely known in the USSR. International recognition came to Aida Andreyevna in connection with her talk at the 3rd International School on Geomagnetic Induction in the Earth, held in Sopron, Hungary, in 1974. The work presented at this talk (Kovtun 1976) has become one of her most widely cited contributions.

In subsequent years (80th and 90th), as high-frequency magnetometers appeared and the theory of processing and interpretation of audio-magnetotelluric data was improved, Aida Andreyevna's interests expanded towards the studies of structure of the Baltic crystalline shield, into the field of investigation of the so-called crustal intermediate conductive layers and crustal anomalies of electrical conductivity (Kovtun 1989a, b). This period includes the discovery of intermediate conductive layers at the depths from 10 to 40 km and their detailed study supported by RFBR grants. One of the most interesting objects for her was the Ladoga anomaly of electrical conductivity, important geotectonic and mineragenic zone at the junction of Archaean and Proterozoic blocks of the Baltic shield. At the turn of the centuries Aida Andreyevna and her team devoted many years of work and a number of papers to this interesting phenomenon. Nevertheless she knew that clear understanding of the nature of this anomaly can be achieved only with the help of all the powerful potential of the modern, XXI century, synchronous complex of magnetovariational and magnetotelluric sounding. So in 2013 Aida Andreyevna became the initiator of a new detailed profile experiment LADOGA "in cooperation with Moscow colleagues". Together with S.A. Vagin, I. L. Vardanyants, N. I. Uspensky and M. Yu. Smirnov she actively participated in the project. Her ideas and advice laid the foundation of the current continuation of the Project at the stage of analysis and interpretation of the data.

Also Aida Andreyevna participated in the deep crustal and upper mantle studies, being one of the leading scientists in the large international BEAR experiment (1998–2002 years) dedicated to the electromagnetic sounding of the entire lithosphere and asthenosphere of the Fennoscandian shield. Her significant contribution to the solution of one of the central problems of the experiment - the determination of the position and geoelectric parameters of the regional asthenosphere - is reflected in a number of Russian and international publications.

Aida Andreyevna enjoyed great respect among leading experts in the field of geoelectrics and hold creative correspondence with Mark Naumovich Berdichevsky, Leonid Lvovich Vanyan, Antal Adam, and Sven Eric Hjelt. She regularly participated in international meetings and conferences. In Fig. 2 you see her participating in the conference in Oulu in 1991.

In 1989, Aida Andreyevna defended a degree of Doctor of Physical and Mathematical Sciences. She is the author and co-author of more than 150 scientific publications, including monographs and three text books. Below, in Fig. 3 two of her most famous books of the time are presented.

Aida Andreyevna became the initiator of the development of the audio magnetotelluric equipment, which allowed the department to obtain new data on the distribution of electrical conductivity on the Baltic Shield. Under her leadership, the work was carried out on the physical and mathematical modeling of electromagnetic fields, which made it possible to improve the technique of magnetotelluric studies in horizontally inhomogeneous media.

Fig. 2. Working moment on the conference in Oulu, 1991 (Finland). From left to right: M. S. Zhdanov, S.-E. Hjelt, V. Yu. Semenov, M. N. Berdichevsky, A. A. Kovtun, A. A. Zhamaletdinov, P. Kaikkonen

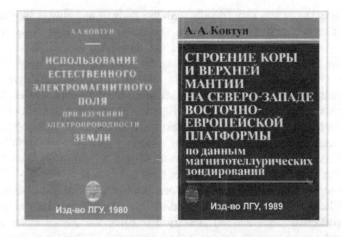

Fig. 3. Textbook and monograph by Aida Kovtun

Under the guidance of Aida Andreyevna, over a dozen of PhD theses have been defended. She was a talented lecturer and taught a number of theoretical courses for graduate students and undergraduates at the Physics Faculty. She also conducted seminars, led magnetotelluric practice, was the head of a large number of term papers, theses and master's theses. Until the very last days, Aida Andreyevna was active in her Department, she directed the electromagnetic section of the "Geocosmos" conference. A fragment of her talk at the conference "Geocosmos-2014" is shown in Fig. 4.

Fig. 4. Speech by A. A. Kovtun at the meeting of the electromagnetic section of the conference "Geocosmos" in 2014, a year before her passing away

Aida Andreyevna was very friendly with the students and always supported young scientists. The talent of the enthusiastic researcher and teacher Aida Andreyevna miraculously combined with the talent of the thoughtful mother of the family, the keeper of the hearth, the teacher of two sons. 56 years she lived side by side with her husband, a well-known scientist, Professor Vladimir Sergeevich Buldyrev, and provided him with comfortable conditions for scientific and pedagogical work. She was his constant partner in work and leisure. Together they conquered mountain peaks, embarked on lengthy kayak trips on the lakes and rivers of the former Soviet Union, made ski routes. Their sons received a good education and work successfully in Russia and abroad. Aida Andreyevna brought up numerous grandchildren who will always remember the kind and wise "grandmother Owl".

But, no matter how full of good deeds and bright events life is, it comes to an end. Husband, Vladimir Sergeevich Buldyrev, died. Aida Andreyevna outlived him by five bief but productive years and on January 15, 2016, she passed away as a Soros professor and a chief research fellow at the Department of Physics of the Earth at the Physics Faculty of St. Petersburg University.

Aida Andreyevna's students and colleagues, engaged in the projects initiated by her, continue her research, remember her and are thankful for her enormous scientific contributions to the theory and practice of the magnetotelluric method. Her memory will live on among her family, many generations of students, and all those who knew her as an outstanding scientist, as well as deeply intelligent person of beautiful spiritual qualities and great personal charisma.

References

Kovtun, A.A.: A constitution of a crust and upper mantle in northwest of the East Europe plateform on data of magnetotelluric soundings. L. Iss. I LIE, 284 p. (1989a)

Kovtun, A.A.: Acta Geodetica, Geophysica, et Montanistica, Academy Sciences Hungary, vol. 11, pp. 333–346 (1976)

Kovtun, A.A.: Structure of the crust and upper mantle in the NW part of Eastern Europe Platform according to the data of magnetotelluric sounding, 284 p. Pusb. LSU, Leningrad (1989b)

Kovtun, A.A.: The use of the natural electromagnetic field in the study of the electrical conductivity of the earth, 195 p. SPb University (1980)

Author Index

© Springer Nature Switzerland AG 2019
A. A. Zhamaletdinov and Y. L. Rebetsky (Eds.): SPS 2018, SPEES, p. 103, 2019.
https://doi.org/10.1007/978-3-030-35906-5

Printed in the United States
By Bookmasters